LIVING WELL WITH

Epilepsy and Other Seizure Disorders

LIVING WELL WITH

Epilepsy and Other Seizure Disorders

An Expert Explains

What You Really

Need to Know

CARL W. BAZIL, M.D., Ph.D.

HarperResource

An Imprint of HarperCollinsPublishers

HarperCollins books may be purchased for educational, business, or sales promotional use. For information please write: Special Markets Department, HarperCollins Publishers Inc., 10 East 53rd Street, New York, NY 10022.

FIRST EDITION

Designed by Joy O'Meara

Library of Congress Cataloging-in-Publication Data

Bazil, Carl W.
   Living well with epilepsy and other seizure disorders : an expert explains what you really need to know / Carl W. Bazil.—1st ed.
      p.   cm.
   Includes index.
   ISBN 0-06-053848-1
   1. Epilepsy—Popular works.   I. Title.

RC372.B39   2004
616.8'53—dc22
                                                            2004047493

04   05   06   07   08   10 9 8 7 6 5 4 3 2 1

MEDICAL DISCLAIMER

This book contains advice and information relating to health care. It is not intended to replace medical advice and should be used to supplement rather than replace regular care by your doctor. It is recommended that you seek your physician's advice before embarking on any medical program or treatment. All efforts have been made to assure the accuracy of the information contained in this book as of the date of publication. The publisher and the author disclaim liability for any medical outcomes that may occur as a result of applying the methods suggested in this book.

PRIVACY DISCLAIMER

The names of the individuals discussed in this book have been changed to protect their privacy.

*To all those living with epilepsy, particularly those whom
I have had the pleasure to treat and who have taught me the most
important thing in disease as in life: it's all about living well.*

CONTENTS

ACKNOWLEDGMENTS

My sincerest thanks to all who have helped me in this long endeavor. First, my many teachers in medicine, pharmacology, neurology, and epilepsy; and particularly my main mentors in epilepsy: Martha Morrell, M.D., and Timothy Pedley, M.D., gave me the foundation on which to build an understanding of health in general and epilepsy in particular. The entire staff of the Columbia Comprehensive Epilepsy Center—physicians, nurses, nurse practitioners, research associates, administrators, and administrative assistants—constantly show me the true meaning of care. Thanks also to all at HarperCollins, particularly Cathy Hemming, Megan Newman, Nick Darrell, and Toni Sciarra, and to the staff at Witherspoon Associates, especially David Forrer and Kim Witherspoon, for their unending patience and guidance in this, my first nonacademic book. Others who helped with reading, advice, or moral support are; Alison Pack, M.D., Linda Leary, M.D., Kelly Parden, David Ebershoff, Jerry Meyer, and my parents and family. The inspiration, motivation, and persistence needed to complete this book came from Eric Price. Finally, the relevance and perspective for everything in this book came from the many people with epilepsy I have met and hopefully helped, particularly those who took the extra time with me to discuss or read this work, and those whose stories appear within its pages.

Introduction

Diagnosis with any medical condition can be worrisome, particularly if you are not well informed about it. Often the uncertainty and misunderstanding can be worse than the disease itself. This is perhaps true of epilepsy more than any other condition. Epilepsy involves the brain, a complicated and poorly understood part of the body that nonetheless is the seat of not only our experiences and our thoughts, but also our sense of self. A condition that affects the brain can therefore be more troubling, perhaps, than one that affects other parts of the body. Most people do not have a good understanding of how the brain works, and therefore of the ways epilepsy could (and more important, could not) affect daily life. In addition, epilepsy has been the subject of prejudice and misunderstanding since the beginning of time. While today people with this sometimes confusing condition are no longer thought to be possessed by demons, and many legal safeguards protect all people in the workplace, the general public is still widely confused about many aspects of this condition. If you have epilepsy, you may find yourself having to educate family, friends, and colleagues while coping with it yourself.

With any medical condition, information is often the most important aspect of care. Armed with a full understanding, people know what to expect, what to look out for—and what not to fear. This book is meant to offer comprehensive information about all aspects of epilepsy. It serves as a guide for those with the condition, and for their family members and friends who would like a more complete understanding of the condition. It is also a guide for professionals or others wanting to know more about this common, sometimes debilitating condition that raises many questions about the human brain: about the source of our thoughts and our fears, and about the meaning of creativity and insight.

Part 1 starts with a description of the normal brain, and how epilepsy can occur. Chapters discuss the different types of seizures and causes of epilepsy, and how epilepsy occurs in specific age groups. Part 2 covers treatment of epilepsy, from traditional to alternative methods. Part 3 delves into the practical aspects of living with epilepsy, including safety, lifestyle, other conditions and their relationship with epilepsy, and childbirth.

I hope that this book contains all of the information needed to fully understand epilepsy and, more important, to live a fulfilling and complete life after diagnosis.

PART ONE

About Epilepsy

1

From Normal Thoughts to Seizures: The Workings of the Brain

The brain is an incredibly complex organ that controls not only what people think and do, but who they are. Ancient philosophers believed that the heart was the seat of the soul, but we now know that it is actually the brain that is the organ of feelings and thoughts, the source of reactions to the world and our interpretations of it. The brain makes us move and speak, see and hear, feel and understand. The brain is also the source of epilepsy, which is probably why epilepsy is such a complicated disease, with manifestations as diverse as the brains from which it arises. Understanding epilepsy, then, begins with an understanding of the workings of the brain.

The brain is an organ of communication, whose job is to process all of the information important to its owner. All of the complicated things that our brain helps us do—translate the spiky shapes of letters and lines on a page into words and thoughts; steer around a pedestrian who suddenly steps into the street; recognize the face of a child and feel overwhelming love for her—consist of electrical and chemical signals that pulse between neurons, which are specialized brain cells. It's sometimes helpful to think of the brain as a living computer. Both brains and computers are made up of individual circuits, both run (more or less) on electricity, and both are capable of retaining and processing information. Both can also malfunction: as a computer screen can briefly freeze, so can a human brain pause during a seizure. But there are differences as well. The human brain can regenerate and heal itself. This is fortunate, because although we can upgrade and buy new computers, we have only one brain throughout life. We sometimes feel that computers do things much more efficiently and easily than our own brains. But no computer can match the creativity and depth of experience that a single human brain has.

This chapter is a brief guided tour through the brain, from the relatively simple individual neuron that is either "on" or "off" to the complex networks of thought and feeling that make us human.

The Brain Cell: The Smallest Unit of Thought

The brain is made up of billions of nerve cells called neurons. Akin to the wires and circuits of the brain, neurons are held together by structural cells, called glia, that protect and insulate the circuits of the brain. Glia are like the insulation on a wire and the backing on which the circuits are mounted. Fundamentally, a neuron (or more accurately a group of neurons) in your brain interprets incoming information, mulls it over, then sends off a signal that results in actions. It is also a neuron (or a group of neurons) that, probably beginning in much the same way, causes all kinds of seizures.

Structurally, a neuron has a cell body that holds most of its workings: the genetic material that determined how it was formed, enzymes that make proteins and transmitter substances, and other machinery for producing energy to run the cell (Figure 1-1). The cell also has "dendrites"—arms of the cell that branch out to the surrounding area, listening to other neurons in the area and taking in-

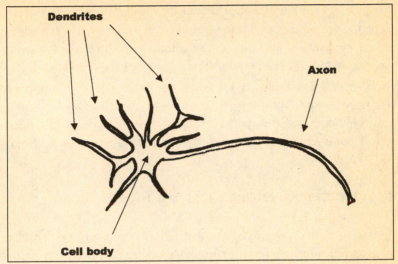

FIGURE 1-1: The Neuron

formation back to the cell body. Some neurons have relatively simple "dendritic trees" that connect with only one or two other neurons; others have extensive branches connecting with dozens or more. Neurons also have an "axon": the output arm of the cell. The axon carries each individual neuron's message to other cells. A neuron, then, looks a little like a scorpion: arms out in all directions, a body that runs the show and decides when to respond, and a stinger that carries a single message to the outside.

The neuron is constantly processing information from other neurons, deciding whether it should "fire" (send a message itself) or stay quiet. When a neuron fires, it releases one or more chemicals called neurotransmitters—serotonin, norepinephrine, epinephrine, and glutamate are examples. Each of these chemicals can either excite—so that the receiving cell gets a buzz and thinks about firing itself—or inhibit—quieting, encouraging the receiving neuron to stay still. When a neuron fires, only other neurons with an appetite for its particular transmitter will be aroused at what are called receptor sites, the place where the neurotransmitter attaches. One neuron may respond to glutamate but not to serotonin. No receptor, and the transmitter bounces off and tries to find another neuron that's listening. So neurons are selective in which other neurons they listen to.

Is a single neuron firing a thought, or can it be a tiny seizure? So far as we understand these, it can be neither. Neuronal firing can only be thought of as a spark. If it falls on ready tinder, a flame can result, but if it falls on water or dust, it fizzles and remains unnoticed.

Neural Networks: From a Spark to a Thought

It may be difficult to imagine how the simple "on-off" system of a neuron can translate into the complex thoughts and reactions of the human brain. The system begins to get more complex when the concept of neural networks comes in. Every brain contains billions of neurons, each of which can talk to one or many other neurons, and each of which listens to other (or the same) neurons. These create hundreds of thousands of neural pathways in a brain, where neurons are connected to other neurons, and there are an infinite number of possible messages. Moreover, neurons can *learn.* As neurons talk and listen to each other, fire and quiet, they can develop relationships among themselves. Some pathways between neurons become grooved with frequent contact, and like a well-worn path through the woods a memory is formed. Other paths become overgrown and hard to find if only used once or twice. The first time any child tries to coordinate the many muscles required to maintain balance while operating a bicycle, it's tough. She must pedal evenly to move forward, shift weight through a turn, and feel the strain in her legs to know when to shift gears. The brain coordinates this process, and as she rides again and again, the same systems are used and refined. They become stronger and easier to access and use each time. Once she masters the skill, the neuromuscular pathways are well paved and clear: get on a bicycle even after many years, and those procedural pathways will waken and guide with an eerie accuracy. The same holds true for other things. Go back to an old neighborhood, one you haven't seen since the age of three or four, and you may find that you mysteriously know the woods' back paths and where the wild tulips grow. The knowledge is engraved in the neural networks, the brain's interstates and country roads. On the other hand, try speaking French after a twenty-year interval. Those roadways have been repaved with the sticky tar of English, hardened now to crust. It will take a lot of work to clear the clogged channels. Thus the simple neuron, firing and quiet, evolves into our living thoughts, skills, and memories over years of use.

Anatomy of the Brain

Understanding of how the brain works must go beyond the neuron to the way the brain grows and develops. The most primitive structures of the brain—concerned with breathing, eating, sexual activity, and emotion—are located at the base of the brain in an area called the brainstem. This part is present in some form in all animals. The newer part of the brain, or "neocortex" (new brain), surrounds the brainstem and is most highly developed in humans. This area is responsible for most of the remainder of brain functions, from complex reasoning to sensation and movement. The neocortex is divided into regions separated mainly by folds in the surface of the brain. Neurons are located mainly in this surface. As humans developed, they needed more and more neurons. To fit all these cells in, rather than growing in size, the brain increased area through many folds in the surface known as sulci. There is a large division between the left and right side of the brain. There is also a large sulcus, called the central sulcus, which runs more or less from ear to ear. These large folds in the brain determine the major regions, or

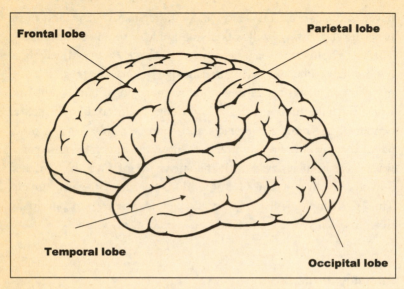

FIGURE 1-2: Lobes of the Brain

"lobes," of the brain (Figure 1-2). The frontal lobes are located in front of the central sulcus, and the parietal lobes behind. Roughly perpendicular to the central sulcus is the Sylvian fissure, which runs backward from the ear and divides the parietal lobe (above it) from the temporal lobes below. The occipital lobes are at the back of the head and are not precisely defined by sulci.

Some structures seem to be in place before a thought ever occurs. For reasons that are not understood, but are probably at least partly genetic, the brain develops particular skills in the same regions in nearly every normal person. The major site for movement is always located just in front of the central sulcus (in the frontal lobe), and each side of the brain controls the opposite side of the body (Figure 1-3). The distribution of control is also nearly identical from person to person: e.g., leg control is deep inside the fissure that separates the hemispheres, hand control is on the side, tongue control is lower down. The finer the control needed, the more brain is involved, so

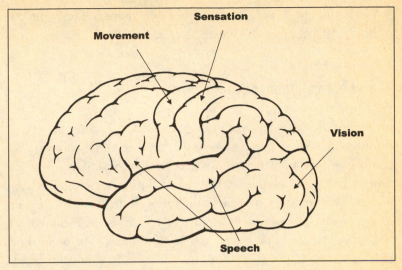

FIGURE 1-3: Functional Areas of the Brain

the hand and the lips use up much more cortical space in the brain than do the back and the neck. The sense of touch is directly behind the motor areas, in the parietal lobes. Vision is located in the back of the head, in the occipital lobes.

Language, a complex human skill, requires several areas. One is located in the frontal lobe, close to the motor areas that control the lips and tongue. This region, known as Broca's area, is primarily concerned with the production of speech rather than its interpretation. Another area, called Wernicke's, is in the temporal lobe and is necessary for understanding speech. A person with a stroke (or a seizure) restricted to Wernicke's area may speak apparently normally, but be completely unable to interpret things that are said to him. If Broca's area is affected, he may understand all that is said, but be incapable of speaking back (except for "automatic" speech such as *yes, no,* or sometimes profanities; these do not seem to require the complex control that Broca's area provides). In nearly all right-handed people, both Wernicke's and Broca's areas are located in the left brain. With left-handed people, language can be on the right, the left, or "mixed," where both sides are used.

The Abnormal Thought: A Seizure

How, then, do these same pathways give rise to a seizure? For one of countless reasons, a group of neurons, or a neuronal pathway, becomes hyperactive. Perhaps those cells were starved of oxygen for a time at birth. Maybe they were bruised during a motorcycle accident, or maybe they were genetically misprogrammed and grew up in the wrong place or with overactive receptors. In many people with epilepsy, the brain looks completely normal. Maybe in these people, the constant reorganization of neurons just randomly made a bad connection, and reinforcement of that abnormal path caused it to grow in strength. In any case, epileptic neurons (or more likely pathways) tend to spurt a lot more transmitter, and far more frequently than they should. As a result, millions of other, normal neurons, caught up in the commotion, begin to fire. If nothing stops it (there are, remember, inhibitory pathways), soon large parts of the brain are triggered and aroused: a single spark, then a flame, and finally a forest fire. In a tonic-clonic (grand mal) seizure, every single neuron is firing simultaneously, a massive lightning storm thundering through the head, so the person cannot think or feel, and all of his muscles are commanded to violently contract, out of control.

Because of the complexity of the brain, epilepsy, to an extent like no other disease, is about individuals: each seizure experience is unique, and each person is touched by it in very different ways. Not only do the different areas of the brain have very different functions, but everyone's brain is wired in a unique manner. A seizure will result in the perception or the response associated with the part of the brain involved. Seizures that begin in the visual areas of the brain begin with the perception of colors and shapes. These can be amorphous and vague: swirls of colors, or a floating, moving object like a huge butterfly. Or they can be incredibly distinct: the face of a deceased relative, or (as in one of my patients) Tweetybird. Other seizures may be accompanied by intense inexplicable feelings of dread; in these cases, the excitation involves the brain's amygdala, a fascinating structure that is involved with emotion, including fear. Sometimes a seizure ripples through the auditory (hearing) regions, and then the person hears clashes, clangs, or lovely strains of music. If the language area is affected, the person may hear, but voices suddenly sound strange and meaningless, "like Charlie Brown's teacher" as several of my patients have explained it. If the gustatory (taste) center is involved, then metal, vinegar, or chocolate can be tasted.

While these examples are fascinating, the majority of people do not experience anything so specific. They may lose consciousness before remembering anything or may only have a vague sense that something is different: a seizure starting in one of the brain's many areas without a function we can easily recognize.

Most seizures start spontaneously, seemingly out of nowhere, but in a few cases they can be brought on by highly specific external cues. In these cases, the normal response to sound, light, or movement must hit an irritated area, and the connection is such that a seizure begins. I once had a patient who loved Billy Joel's music, then began to have seizures every time she heard "Piano Man." In a well-publicized case, a woman had seizures specifically triggered by newscaster Mary Hart's voice; in another, hundreds of Japanese children had seizures from watching a single episode of *Pokémon*. Although unusual, this sort of reaction should not be surprising; after all, the irritated brain involved in seizures remains intimately connected to the normal surrounding areas. So just as a seizure can produce sensations that seem real, so can a real sensation occasionally trigger a seizure.

Depending upon the individual and the meaning she attaches to these strange states, seizures can have very different impacts on the person who experiences them. They can be seen as a gift, a curse, a bother, a bore. But all impact the person's life in ways that can be confusing, terrifying, or even dangerous, particularly in the absence of adequate knowledge of what is happening. Seizures begin through the same machinery that produces our thoughts and feelings; some of the small ones may even be difficult to distinguish from normal feelings. Most seizures, however, are at least problematic and frequently disruptive. The next chapter describes specific seizure types in more detail.

2

Tickling the Brain in Many Ways: The Many Types of Seizures

One of the most common misperceptions about epilepsy is that it is a single disease. In fact, epilepsy actually includes a wide variety of conditions that have one thing in common: the brain malfunctions spontaneously, then returns to normal. Usually this means *entirely* normal: I have patients who are doctors, nurses, actors, artists, and writers, all successful in their fields. This process of normal functioning most of the time, but interrupted by seizures, sounds rather simple but is in fact complicated. Epilepsy comes about for a variety of reasons, from conditions that change the structure of the brain (like a stroke) to genetic conditions that alter the way the brain reacts. The latter are mostly "generalized epilepsies" and will be described in detail in the second part of this chapter. Generalized seizures are one of the two broad divisions in seizure type. They are called generalized because they seem to begin in all areas of the brain at once. The other seizure type, and the most common, is "partial seizures": seizures that begin in a discrete area of the brain and can then spread to involve other areas or the entire brain. Because each part of the brain has distinct functions, partial seizures are incredibly variable in different people.

Many people think that partial seizures are somehow less severe or serious than generalized seizures. This is not necessarily true; the terms have only to do with how the seizure starts, and not with severity. Arguably the most "mild" seizures are "absence seizures," consisting only of a brief staring; however, this is a generalized-seizure type. It is also confusing that a grand mal (generalized tonic-clonic) seizure can be either partial or generalized; it depends only on whether it starts small and spreads (partial, and known as a secondarily generalized tonic-clonic seizure) or whether it seems to start all over the brain (called a primary generalized tonic-clonic seizure). It may in fact be impossible for even a well-trained neurologist to distinguish a primary from a secondarily generalized seizure as she watches them.

The Many Types of Partial Seizures

As mentioned above, although one seizure type is called partial, that does not mean that they any less severe, debilitating, or dramatic than generalized seizures. The term refers only to the discrete onset of each seizure: partial seizures begin in one part of the brain.

A commonly used scale ranks the severity of partial seizures: they may be simple, complex, or secondarily generalized.

A "simple partial" seizure means that the seizure is small enough that the person does not experience any confusion or loss of awareness. For instance, a patient can have a seizure that begins with a thumb jerking back and forth. She is fully aware that the thumb is misbehaving and remains aware as the other fingers or the arm join in. This is a simple partial motor seizure. Some simple partial seizures that do not involve movement are called auras. Most people (including many physicians) think of an aura as a warning rather than an actual seizure, but in reality it is a small seizure. Common auras are a feeling of dread or doom, a feeling of déjà vu, or a sudden foul smell. An aura can be distinguished from other simple partial seizures because it is totally subjective: the person experiencing it is the only one who can feel or see that it is happening.

A "complex partial" seizure includes some degree of confusion. A simple partial seizure can become complex by spreading to involve more of the brain. Probably, there has to be some involvement of both sides of the brain before awareness is altered. During a complex partial seizure, a person can appear fairly normal: he may continue to stand, walk, or even talk, but he is partly or completely unaware of what is happening around him. Afterward, he may have no memory of what transpired during the seizure.

A "secondarily generalized" seizure starts in one area of the brain, but slowly or quickly spreads to involve the entire brain. It may appear identical to a "primary generalized" seizure, described above. Or the jerking thumb can become a jerking hand, arm, arm and leg, then start on the other side (originally described in the nineteenth century by a well-known neurologist named Hughlings Jackson, this "traveling" seizure is called a Jacksonian march). The end result of both primary and secondarily generalized seizures is that the person becomes stiff in all extremities, ("tonic") followed by rhythmic ("clonic") jerking movements of the entire body. With a secondarily generalized seizure, the person may not remember the first part, when the seizure is small. In some cases, the "aura" will give enough warning that the person can get to a place of safety before losing consciousness (but typically loss of consciousness occurs within no more than about thirty seconds).

Seizures Vary by Their Region of Onset

In many cases, it is possible to tell where in the brain a seizure starts simply by what it looks or feels like. Although the exact function of vast areas of the brain is unknown, many areas are also known to govern specific functions in all people with normal brains. Much of this information was determined early in the twentieth century by a surgeon, Walter Penfield. When he did brain surgery, he kept people awake so that he could ask for their responses. This may sound frightening and painful, but strangely, the brain does not perceive at all any injury to itself. So if the scalp and the coverings of the brain (called dura mater) are anesthetized using local anesthesia (the same as is used for most dental surgery), the surgeon can cut into the brain and the person feels nothing.

Once the surface of the brain was exposed, Penfield used an electrical wire to touch areas he thought needed to be removed. If nothing happened, he could safely assume that the person would be okay if that portion was removed. However, the person might say, "I see a yellow orchid on the left," and Penfield would know that area was needed for visual processing. Or the right arm might contract, and he knew that the area controlled the person's movements. Through experience with many people, Penfield generated a "map" of the functions assigned to various parts of the normal brain. Since that time, Penfield's maps have been refined by many others. The map varies some between people, particularly if areas of the brain are damaged early in life; in this case, other areas often compensate for lost functions. But the descriptions below are accurate for the vast majority of people. Language is a key area that differs in location among people. As I've mentioned, right-handed people almost always have language functions localized in the left hemisphere, but in left-handed people it can be in the left, the right, or both hemispheres.

Frontal Lobe Partial Seizures

The frontal lobe is usually associated with movement, and most seizures that begin here have motion as a primary manifestation (Figure 2-1). The "primary motor area" is a strip of brain at the back of the frontal lobe, bordering the parietal lobe. Seizures that begin here usually consist of repeated, strong jerking beginning on one side in the face, arm, or leg. If the seizure spreads, the rest of the

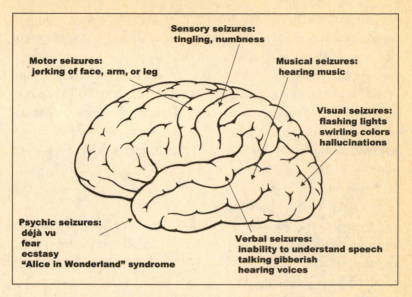

Sensory seizures:
tingling, numbness

Motor seizures:
jerking of face, arm, or leg

Musical seizures:
hearing music

Visual seizures:
flashing lights
swirling colors
hallucinations

Psychic seizures:
déjà vu
fear
ecstasy
"Alice in Wonderland" syndrome

Verbal seizures:
inability to understand speech
talking gibberish
hearing voices

FIGURE 2-1: Location of Various Partial Seizure Types

side and the opposite side can progressively become involved (a Jacksonian march, as mentioned earlier). Seizures also begin in the "supplementary motor area"—located in front of the primary motor strip and deep in the cleft between the hemispheres. These seizures also involve a lot of movement, but it is more chaotic. Sometimes the person will look as if he is riding a bicycle or swimming. At other times, the movements will look completely random. For reasons that are not known, frontal lobe seizures begin commonly or exclusively during sleep and are usually briefer than other partial seizures. One patient of mine had simple partial motor seizures exclusively in his sleep, consisting of awakening, grimacing, and tensing of all muscles. He was completely aware during the seizures. Although they lasted only a few seconds, they would occur many times per night, and he had great difficulty functioning the following day due to his disrupted sleep cycles.

The motor areas of the frontal lobe actually comprise only a small portion of its area. The frontal lobes are by far the largest of the lobes of the brain, but much of the function of the remainder is less distinct. The remainder of the frontal lobes are largely concerned with reasoning, social skills, and higher thought. People may very well first experience some type of change that they cannot describe, but as the seizure spreads, the involvement of the motor areas will usually finally result in the kinds of motor seizure described above.

Parietal Lobe Seizures

Seizures in this area are not terribly common. The most distinct function of the parietal lobe is sensation, and this is the most obvious manifestation of seizures in the parietal lobe. Juan, for instance, suddenly feels a tingling, as if his left arm were falling asleep. The skin puckers up in "gooseflesh," as if he had gotten a sudden chill. Even after many years of experiencing these early sensations of seizure, Juan still shakes his arm when this happens, hoping to "wake it up," even though this never works. After about a minute, the sensation starts to subside. In the past he would know that he needed to sit down immediately when he felt these sensations, because he would likely pass out. Since he started medication about three years ago, Juan no longer has altered consciousness, although the milder seizures persist.

Parietal lobe seizures almost always have some sort of sensory component. Most commonly it is described as a tingling or "electric" feeling, but sometimes it feels more like burning or numbness. The primary sensory area is a strip of brain in the front of the parietal lobe, just behind the primary motor strip in the frontal lobe. When parietal lobe seizures spread, they will therefore commonly hit the motor strip, and the tingling will be replaced by uncontrollable shaking. If the seizure spreads backward, into the occipital lobe, visual hallucinations can begin.

Occipital Lobe Seizures

At forty-one years of age, David began a transformation in his painting style. For a decade he had created canvases covered with shifting shapes, abstract blocks of color that seemed to shimmer and move across the space. As his painting interests shifted to more representational views, he began to sense something strange: the abstract images of his paintings seemed not to want to leave his mind. He began noticing them: fleeting sheets of crimson and deep violet, bizarre amorphous creatures appearing from nowhere, always on the left side, lasting at most a minute and often no more than a few seconds. They brought with them a calm, pleasurable feeling. One afternoon, as he sat on the porch of his seaside summer house, the colors came harder and brighter. They did not flit in and out of sight; they rolled over him and his pleasure intensified. Then calm turned to terror as his left hand clenched into a claw; his arm spasmed, and he fell down in what was now, clearly, an epileptic seizure. His wife was startled by her husband's violent outcry as his chest muscles contracted and air was forced from his lungs. She found him on the ground, shaking violently, with blood oozing from his lacerated tongue as the convulsion gave way to quiet exhaustion.

The occipital lobe is primarily concerned with vision. In the very back of the brain, an area called the calcarine cortex contains the primary visual area. This is the first stop for image recognition, but there is no real processing. This area sees the world as a series of dots and circles, as when a picture is blown up so big that you see variation in pattern or color, but no object. Seizures that begin here are visual, but abstract: a splotch of color, a blinking light, a swirling mass. Just in front of the calcarine cortex is the "visual association area," where the simple dots are processed into images and are associated to the brain's known world. The dots will form a shape with limbs; two circles above another with a tongue will become a face; a shape with a tail will be recognized as an animal; and finally the neighbor's cat is recognized. A message to the language area, and his name, Simon, echoes in your head. Another connection and you tense up, remembering last week when Simon took an angry swipe at your nose.

A seizure starting in the visual association areas will be precise. One patient of mine saw an image of Tweetybird; another saw the face of an otherwise forgotten third-grade teacher. As with David, people who have limited seizures in the occipital lobe can find them more amusing than troublesome, until they become the harbingers of a more serious attack. People with purely visual hallucinations as seizures can also be misdiagnosed as having other diseases, especially migraine, where attacks commonly begin with changes in vision.

Temporal Lobe Partial Seizures

The temporal lobes are the most frequent site of onset for partial seizures. The most concrete functions of the temporal lobes concern memory and language, so these can be affected by seizures. Deep within the temporal lobes are primitive parts of the brain concerned with emotion and memory. For some reason, these structures, called the hippocampus and the amygdala (from the Greek for "sea horse" and "almond," for their respective shapes as seen by early anatomists), are particularly prone to develop seizures. Fear or ecstasy, changes in perception of time, or feelings of unexplained familiarity (déjà vu) or inappropriate unfamiliarity are therefore common manifestations of seizures from the temporal region.

Gloria would suddenly have an overwhelming feeling of déjà vu, as if what was happening had all happened before. The feeling was not frightening in itself, but she knew that she was likely to lose consciousness within a moment or so, so she would find a place to sit (or simply sit on the floor). Sometimes, the feeling would just go away. More often, though, her perception of voices would change next. If someone was speaking to her, the words would sound like nonsense. She would try to tell them to speak English, to stop mumbling, but no words would come from her lips. She would then start to absently rub her cheek. Then, as suddenly as it had started, the seizure would stop and she would find herself sitting quietly, sometimes with a stranger or a friend looking at her with concern.

Gloria's seizures started in the inner part of the temporal lobe, in the hippocampus. This is an ancient structure, common to all mammals. It is needed to form new memories. In Gloria's seizure pattern, her hippocampus gets an erroneous signal that her current reality is an old memory rather than a new one, and a strange sense of familiarity comes over her. The signal grows to involve her nearby language areas, which, engulfed in the abnormal brainstorm, can no longer process the sounds of speech. In other people, the seizure can "tickle" the areas of emotion, triggering a sense of unfathomable dread or sublime ecstasy. Distortions of time or space can happen: like Alice in Wonderland, the person may feel that she is becoming smaller and smaller or larger and larger. Time stands still or races forward such that a month seems to have passed in a second. Some people, rather than having speech impeded, begin to repeat a phrase over and over without knowing what they are saying.

As I've mentioned, nearly all people have language on only one side of the brain; this side is called dominant and almost always is the same side that controls the dominant hand. The other side is usually involved with music (I have always found it interesting that the brain seems to find each of these equally worthy of their own hemisphere for appreciation and interpretation). People with damage to their dominant temporal lobe may be unable to read words, but have no problem reading music. Even more interestingly, stroke patients who have totally lost the ability to speak due to damage to the dominant hemisphere are usually able to sing words—song lyrics must become inextricably coded into the nondominant, musical hemisphere. A nondominant temporal lobe seizure can begin with the illusion of music: a tune, usually familiar, playing in the head. It can be a hymn from childhood, Handel's Water Music, or the Supremes singing "Baby Love." The reverse also can be true; as I mentioned, for one of my patients, hearing the piano introduction of Billy Joel's "Piano Man" caused a seizure by tickling the vulnerable temporal lobe path. As with all seizures, one of the keys to recognizing the feeling as abnormal is what we call stereotypy: it is the same thing every time.

As you can see, there is a great diversity of partial seizures. Keep in mind, though, that many are not nearly as easily described. People frequently have no memory of how a seizure begins. Maybe they see an image of Marilyn Monroe at the start, but the effect on the hippocampus is such that they never remember it later. Or perhaps the seizure starts in a quiet area, so the feeling can't really be described. A detailed description of what happens particularly in small, simple partial seizures can, however, be helpful in determining where in the brain the problem lies. For instance, it can cause doctors directing diagnostic tests, such as MRI pictures of the brain, to look at certain areas more carefully.

Generalized Seizures

The term *generalized seizures* confuses many people. While the term seems to imply that the attack is larger or more dramatic than a partial seizure, this is not always the case. In generalized seizures, the entire brain seems to be involved all at once, rather than the seizure beginning in a particular spot (as in partial seizures). Strangely, however, involvement of the entire brain can carry very different symptoms, from the subtle, easily missed pause of a brief absence seizure to the dramatic convulsions of a primary generalized tonic-clonic seizure. Below, I'll profile the various types of generalized seizures and will briefly mention treatment, which we'll cover more extensively in chapter 9. Next, "syndromes" of generalized seizures are described. Many generalized epilepsy syndromes include two or more types of generalized seizures occurring together. Finally, I'll discuss how it is that so many different seizure types can arise from a seemingly similar phenomenon: the appearance of a sudden seizure discharge involving all of the brain at once.

Absence Seizures

Absence seizures are colloquially known as petit mal, although that term is often misused by patients, the press, and even many physicians. A classic absence seizure consists of a momentary pause in action by the person. This can be accompanied by increased blinking, but no other changes will be apparent even to a trained observer. During the seizure, lasting at most a few seconds, the person will not be aware of what is going on around her. But she will not fall down or shake and will likely even continue to hold on to whatever may be in her hand. In a quiet environment, she may be totally unaware that anything abnormal has occurred. If in the middle of a conversation, she may notice that she missed something. Similarly, a friend or family member may not note a problem unless, for instance, the person stops speaking in the middle of a sentence. Typically, even if that occurs, she will resume talking coherently immediately after the seizure stops. Absence seizures can remain undiagnosed for years. People sometimes wonder whether lack of attention or daydreaming is a seizure. In fact, the distinction can be difficult without an electroencephalogram (EEG), a measurement of brain waves commonly used in diagnosing epilepsy.

If an EEG is obtained during an absence seizure, it will show a specific pattern of brain activity known as three per second spike-wave: neurons all fire together during the seizure at that specific frequency, overwhelming normal brain-wave patterns. Once the seizure stops, activity immediately returns to normal. For a more complete description of the EEG, see chapter 7.

If an absence seizure lasts longer than a few seconds, other manifestations may become apparent. The person may purse her lips or make smacking noises. Arms or legs may subtly twitch. If it goes on longer, she could fall, although this is still usually not the sudden, violent drop seen in other seizure types. Whenever these other manifestations are present, the seizure may be called an atypical absence because a typical absence seizure is brief and limited.

Tonic Seizures

Tone refers to the stiffness of a muscle at rest. Muscles are never completely inactive; rather, a continuous, low-level contraction occurs at all times (although this decreases normally during sleep). Normal joints are not floppy like those of a rag doll, but rather slightly resist movement. This helps in retaining posture and also keeps the muscles ready for movement at all times.

In a tonic seizure, part or all of the body's muscles suddenly become stiff in a single forceful contraction. If the person is standing, this can result in a sudden pitching forward. Because the arms are also locked in the tonic contraction, they can't break the fall and there is a real chance of injury, particularly traumatic injury to the head. The seizures usually only last a second or two, but this is long enough to cause a potentially serious fall. People who have uncontrolled tonic seizures may therefore need to wear a helmet to prevent serious injury.

Atonic Seizures

This seizure type is actually the reverse of a tonic seizure. It consists of all muscles suddenly losing tone (along with all voluntary control of muscles). In the absence of tone, the person becomes like a rag doll and will suddenly collapse. Strangely, the seizure can appear similar or identical to a tonic seizure, with similar risk of injuries especially to the head, making distinction between tonic and atonic seizures sometimes difficult. Fortunately, both respond to the same types of treatment.

Clonic Seizures

Clonic seizures involve muscles as well, but rather than becoming stiff or relaxing completely, the muscles contract and relax, over and over. A generalized clonic seizure will therefore consist of jerking and relaxing, jerking and relaxing, about once every second. Partial seizures can have similar movements but only involve part of the body—for example, an arm jerking over and over. Generalized clonic seizures are unusual, unless combined with another seizure type (such as tonic-clonic, below).

Tonic-Clonic Seizures

These classic "grand mal" seizures are what most people think of when they hear *seizure* and are also how epileptic seizures are most commonly depicted by the media. First muscle tone increases tremendously throughout the body (the "tonic" component). The person will fall, and her back may arch with the force of the muscle contraction. The forceful contraction also involves the chest and diaphragm, such that a loud outcry usually occurs as air is forced from the lungs. This usually lasts a few seconds, then the limbs begin to twitch rhythmically and together (the "clonic" component). The twitching may become faster and faster before suddenly stopping, usually within a minute. There may be foaming at the mouth as secretions are forced out by the sheer force of the contracting abdominal and chest muscles. Similarly, urine or feces may be forced out. The seizure is followed by extreme exhaustion, as the overworked brain shuts off and the person lies still, but unresponsive to the world.

As with tonic or atonic seizures, people can fall during generalized tonic-clonic seizures. Injury is not uncommon, although fortunately this is usually no more than bruising or scraping. Because of forceful contraction of the jaw muscles, the tongue can be bitten severely (but heals quickly). A common misperception is that people can "swallow their tongue" having this type of seizure. This is impossible; the tongue cannot be swallowed. It is frequently bitten during a generalized tonic-clonic seizure, but putting objects in the mouth of someone having a seizure is a bad idea and is not likely to prevent this. More likely results are broken teeth in the person who had the seizure, or seriously bitten fingers of the well-meaning caregiver.

Myoclonic Seizures

Myoclonus, like clonus, consists of a sudden jerking of the muscles. It differs mainly in the speed of the jerking and relaxation; with myoclonus, both are faster and take place in a fraction of a second. Most people have experienced normal myoclonic jerks as they fall asleep, where the entire body suddenly jerks. This has no relationship to seizures, but the movement is similar to that which occurs in a myoclonic seizure. Myoclonic seizures may not be terribly debilitating, and they can be misinterpreted as simple clumsiness. Sometimes, larger myoclonic seizures can result in dropping things or even falling.

Epilepsy Syndromes

Many generalized seizure types occur as a part of a particular disease, or syndrome. Diagnosing the epilepsy syndrome, as opposed to the seizure itself, is important because it helps to define the cause of the epilepsy, particular treatments that may work better than others, and the prognosis (or expected long-term course) of the disease. Note that a given seizure type, such as generalized tonic-clonic, can occur in several epilepsy syndromes. Also keep in mind that many people's symptoms do not precisely fit a known syndrome, and that there may be overlap between syndromes (people can have characteristics of more than one).

Syndromes that occur primarily or exclusively in childhood are described in chapter 4, "Children Are Different."

Juvenile Myoclonic Epilepsy (JME)

Juvenile myoclonic epilepsy is one of the more common epilepsy syndromes, occurring in about one in ten of people with generalized epilepsy. Unfortunately the diagnosis is often missed, potentially resulting in suboptimal treatment. People with JME usually begin to have seizures during adolescence (hence *juvenile* in the name). They have at least two seizure types: generalized tonic-clonic and myoclonic. The myoclonic seizures can be relatively mild and are often overlooked particularly because the generalized tonic-clonic seizures are so much more dramatic. Mild, jerky myoclonus can be confused with simple clumsiness. Myoclonus is usually worse in the morning, and a common story is that the person frequently and inexplicably spills coffee or milk at breakfast. The generalized tonic-clonic seizures also tend to happen more often just after awakening, particularly if the person has been sleep-deprived. People with JME sometimes have absence seizures as well; these can also be overlooked particularly in the presence of grand mal seizures.

JME is particularly important to diagnose for two reasons. First, it is usually easy to control provided the correct medication is used. Valproic acid is the most common treatment, and sometimes a low dose is sufficient. Other "broad-spectrum" drugs, such as lamotrigine, topiramate, and zonisamide are also used. Drugs such as phenytoin, carbamazepine, oxcarbazepine, and gabapentin can control the generalized tonic-clonic seizures and are frequently prescribed, particularly when the myoclonus is overlooked. However, any of these can make myoclonus (or absences) worse. Lamotrigine can also worsen myoclonus in a small subset of individuals.

The other reason JME is important to diagnose is that lifetime treatment is usually required. Whereas in many cases it makes sense to try discontinuation of medication after several years of seizure freedom, people with JME will almost invariably have seizures if this is attempted. So, while JME is usually easy to control fully, lifetime treatment is typically required, and therefore it's particularly important to choose a medication with minimal long-term adverse effects.

Awakening Grand Mal Epilepsy

This syndrome is similar to JME in that generalized tonic-clonic seizures tend to occur upon awakening or shortly afterward, but differs in that no other seizure types are present. Seizures tend to start in adolescence or early adulthood. They are usually easily controlled with most medications.

Because these seizures occur only at specific times of the day, people with this type of epilepsy should be particularly cautious and avoid driving shortly after waking up. As in JME, seizures can be more likely if the person has been sleep-deprived.

Lennox-Gastaut Syndrome

This syndrome includes many different seizure types, typically a combination of partial, tonic, atonic, atypical absence, and/or generalized tonic-clonic. People with this syndrome usually start having seizures in infancy, with infantile spasms (West syndrome). In childhood, the other seizures develop, and the particular combination can change from year to year. Because of the many seizure types, broad-spectrum anticonvulsants are usually best (valproic acid, lamotrigine, topiramate, zonisamide, or felbamate). This syndrome is usually difficult to control completely, and two or more medications may be required. Often the most troublesome seizures are tonic or atonic, resulting in frequent injury. If these cannot be completely controlled with medication, surgery can be helpful in limiting the number and severity of seizures.

People with Lennox-Gastaut syndrome usually have other neurological problems as well. Almost all have below-normal intelligence, and many have mental retardation. It is not clear whether this arises from repeated seizures over many years, or whether brain damage from unknown causes results in seizures.

Generalized Symptomatic Epilepsy

All of the above seizure syndromes, with the exception of the Lennox-Gastaut syndrome, probably have some hereditary basis. The actual problem has not been identified, but presumably some system, or systems, in the brain are susceptible to overactivity, and the person was born with that susceptibility. It is also not known why, if these syndromes are hereditary, many (such as JME) do not manifest themselves until later in life. This likely has something to do with the overall development of the brain. It is known that some transmitters that excite the brain in a child become inhibitory later, theoretically changing pathways such that an imbalance (and therefore a propensity for seizures) develops. Also, the underlying problem may not be capable of producing seizures until certain connections and pathways are well established in adolescence.

People with generalized symptomatic epilepsy may have identical-appearing seizures—tonic, atonic, atypical absence, generalized tonic-clonic—but their epilepsy arises from an event that damages the brain. These individuals are born normal, but strokes, infections, trauma, or a combination of these somehow creates a malfunction in the brain. The treatments are similar to those used in people with Lennox-Gastaut syndrome.

How Do Generalized Seizures Occur?

The exact mechanism by which a generalized seizure happens is not known. On an electroencephalogram (EEG), which measures the electrical activity of the brain, when she is not having a seizure, the brain activity of a person with generalized seizures is largely the same as with anyone: chaotic, with different electrical rhythms that are constantly changing. During a generalized seizure, the EEG becomes simple, with a pattern of electrical activity that repeats rhythmically and monotonously. The normal varied activity of the brain is overwhelmed by vast numbers of neurons firing in unison.

How is it that a similar pattern of activity, seemingly involving the entire brain, can result in these very different seizures—from the simple staring of absence seizures to dramatic tonic-clonic convulsions? Probably different networks of neurons are involved in the various generalized seizure types. Even though it looks as if the entire brain is engulfed in the seizure, in many generalized seizures, it clearly is not. A person with an absence seizure is able to continue standing, or even to hold something—indicating that the part of the brain that controls these functions is relatively unaffected. Tonic, myoclonic, and atonic seizures probably involve other networks that govern a relatively small part of the brain, and in different regions from those in absence seizures. Generalized tonic-clonic (grand mal) seizures probably involve a lot more of the brain, as these are much more violent and require a longer period of recovery. Still, though, all of the brain is probably not engulfed in the seizure.

Table 2-1 opposite summarizes the various types of seizures, and Figure 2-1 (on page 19) shows the location of partial-seizure types in the brain.

TABLE 2-1: SEIZURE TYPES

Seizure Type	Other Names	Description
Partial		
Frontal	Major motor	Uncontrolled movement (usually one side initially)
Temporal	Psychomotor	Déjà vu, language disturbance, music, quiet staring
Parietal		Tingling, numbness
Occipital		Flashing lights, colors, visual hallucinations
Generalized tonic-clonic (secondary)	Grand mal	Any of the above, leading into tonic followed by clonic movements of the entire body
Generalized		
Absence	Petit mal	Brief staring
Atypical absence		Prolonged staring with drooling, blinking, or chewing
Atonic	Drop attack	Loss of muscle tone, falling
Tonic		Tensing of all muscles, falling
Clonic		Rhythmic jerking of all muscles
Myoclonic		Sudden twitching of muscles
Generalized tonic-clonic (primary)	Grand mal	Tonic, followed by clonic movements of the entire body

3

How Does It Happen?
Causes of Epilepsy

The brain is the most complicated organ in the human body, but paradoxically, its responses to injury are simple. No matter what nasty things the body or the external environment do to the brain—poisonous venoms, dangerous infections, a life-threatening lack of oxygen—it responds in really only two ways to tell the person (and the doctor) something is wrong. One way is to simply shut off: the person faints. Everything in the brain stops completely until the injury corrects itself. With the most common kind of fainting, the blood supply might already have been limited because of, perhaps, being in a warm room. Blood then tends to go toward the skin to cool the body. Add that to standing, making it harder for the heart to pump blood against gravity into the brain. Finally, there's a paradoxical reaction when the person becomes frightened, further slowing the heart. Result: no blood (and no oxygen) to the brain, so the person passes out cold and falls to the floor. Voilà! The heart is now at the same level as the brain, no more fighting gravity, the brain is happy again because of the oxygen-rich blood charging through it, and the person wakes up.

The other possible reaction is also rather simple in principle, although incredibly rich in its spectrum of possible manifestations: a seizure. Things that irritate the brain, particularly those that irritate a particular part of the brain, can cause neurons to fire erratically. If the cause is external—too much cocaine, a viral encephalitis, or a really hard blow from a baseball bat—then the brain itself is normal, and this is not epilepsy but a provoked seizure. Sometimes, though, a constant irritant can give rise to repetitive, spontaneous seizures (epilepsy). In most people with epilepsy, a definite cause is never found, although there is much speculation of how the abnormal pattern arises.

For Marissa, the cause of her seizures was found, although it took nearly twenty years. When she was a young girl her seizures began in an unusual way, as a kind of a game, initially known only to her. Growing up in Colombia, Marissa was always playful, especially with words. She loved tunes and nursery rhymes. Although she began learning English at the age of three, she preferred her native Spanish because of its tuneful, happy-sounding tones.

At around age five, Marissa discovered a fun game. If she repeated a little phase in Spanish, *La chiquita niña, niña y madre. Chiquita niña, niña y madre*, the words would begin to take on a life of their own. They would seem to repeat themselves without her trying and would twist and turn about themselves. It was great fun, and she could amuse herself for hours—on the sunporch, in the car, and especially during the heavy boredom of grade school. Not only would she be amused, but she got an intensely pleasant sensation, almost as good as eating chocolate ice cream. Sometimes adults scolded her because she wasn't paying attention. That wasn't fair, because she never heard a thing they said.

As Marissa grew older, the game grew less interesting but would sometimes start without her wanting it to. She could be in a really exciting class in art, trying to absorb all of the beauty and depth of Renaissance paintings, when the stupid phrase would start. She could usually stop it, but not always. During adolescence, like most girls, she spent endless hours talking with friends about boys, about exotic travel, about fashion. At about that time a new kind of game, an unwelcome one, began. Not only would her own little phrase invade her mind, but the phrases of others began to do the same thing. Simple phrases—"Want to go to that new restaurant, Marissa?"—twisted and turned—"restaurant Marissa, new restaurant Marissa, go new restaurant"—and she would have trouble answering. People thought her absentminded, slow, a daydreamer. But she was pretty and smart, and she still excelled.

At seventeen, she was driving to the beach one day with her friends. Again, the phrases started coming—this one about the black dress another friend had worn to the discotheque the night before. "Swirling black dress, funny black dress, swirling black, swirling funny." She sighed and tried to fight it. The pleasure was still there, but this was the wrong time and she grew angry and tired of these things. "Funny black, swirling dress," it kept going and going. She vaguely thought her friends were speaking to her but she felt herself fading away. This time, for the first time, she became completely unconscious. Her limbs stiffened, and she cried out loud. Then she had jerking movements of her arms and legs for a minute, followed by complete exhaustion.

Of course, she had been having seizures all along, but this time it had progressed into a convulsion, a secondarily generalized seizure. When the phrases repeated themselves in her head, this was actually a simple partial seizure. It was unusual because she could actually induce it herself—few people can do this. Such seizures would likely begin in or near the language area. When she became "absentminded" during these episodes, she was actually experiencing a slightly larger seizure, a complex partial one. She was not absentminded, but partly unconscious. Because all of her seizures were relatively subtle, she never consulted a doctor until the grand mal convulsion occurred.

Marissa had the usual tests available in Colombia at the time: a CT scan of her brain and an electroencephalogram. The first was normal, the second showed abnormal spikes of high-voltage activity in her left temporal lobe, confirming that she had epilepsy, and she was prescribed phenytoin. The little spells improved, though they still happened from time to time. These were not so bad. She didn't have any more convulsions until three days after she began another brand of phenytoin (the pharmacist was out of the one she usually took). Four years passed and she had a total of three more convulsions, all either when she could not obtain her usual brand of phenytoin or when she became sleep-deprived.

Marissa had always been interested in music, and she came to New York to study at the Juilliard School. Another neurologist who works at my center examined her for her epilepsy and ordered an MRI scan of her brain. This test shows much more detail than a CT scan, and Marissa had never had one. Although the chances of finding anything serious at that point were small—after nearly twenty years, a nasty tumor would have revealed itself—still, one MRI would confirm no serious problems and is now performed in most places when epilepsy is suspected.

It turned out Marissa had a lesion, or abnormal area, in her left temporal lobe. In her case the lesion was a cavernous angioma: a small, abnormal collection of blood vessels that had probably been with her since birth. Bleeding of these is quite common. Although the brain needs a greater blood flow than any other organ, blood is, strangely, extremely irritating to the brain when it comes in direct contact with brain cells (usually blood is restricted to blood vessels). Almost certainly, this malformation had bled at least once during her childhood. The resulting irritation set up a seizure focus close to her language area. Initially, that funny childhood phrase must have somehow kicked off just the right combination of nerve pathways to start a small seizure, wherein the words repeated themselves, and if the seizure spread a bit, the entire language area was affected until she wasn't understanding anything being said. As time went on, the seizure pathway had the power to start by itself, or with different words. Maybe there had been another small bleed. In any case, the cause of her seizures was now known.

What are the causes of epilepsy? There are many, and these differ somewhat depending on the age at which seizures begin. In fact, epilepsy most often starts at the extremes of life—in the very young or the very old—although it can begin at any age. The potential for injuries and insults that can later cause epilepsy really begins at conception, with the completion of a person's genome (our genetic material). From that time on, the many things in our lives that affect the brain also have the potential to cause seizures to begin. The remainder of this chapter describes common situations in which epilepsy can develop. These causes are summarized along with their relative frequency in Figure 3-1 at the end of this chapter.

Genetic Causes of Epilepsy

Conception: Twenty-three chromosomes from the mother join twenty-three chromosomes from the father. At this moment, the genetic code of a human being is formed and will remain throughout that person's life. Your genetic background determines more than the color of your eyes, your height, and your potential for developing certain diseases such as diabetes. It also determines all of the chemicals and structures that make up your brain. While these are similar in different people, they vary in certain enzymes and receptors. Think, for example, about the nose. Everyone has one, and everyone uses it to breathe. Noses vary, however, in structure. This makes them look somewhat different and also changes the amount of air that can be moved through the nose. Most of the time, the variation in air movement between people is not a problem, but occasionally someone's nose is so small or formed in such a way that the person has trouble breathing, and it needs to be corrected. Perhaps in a similar way, genetics determines all of the structures needed to build neurons, synthesize transmitters, and release them when a neuron is active. The slight variations in these from person to person don't usually result in a problem, but sometimes a combination of factors make neurons overactive and cause seizures. Perhaps a variation in a transmitter is not enough to cause epilepsy, but if it occurs in the same person as a change in the receptor, enough overactivity may result to cause spontaneous seizures.

A number of types of epilepsy are genetic, and thus the potential for seizures is present from the time of conception. Not surprisingly, most of these people begin to have seizures during childhood. Most of these, as far as we know, are generalized epilepsies. Perhaps a mutation in the gene that creates the sodium channel (part of a neuron required for firing) makes it easier for neuronal firing to get out of control. Or maybe a receptor for glutamate, one of the excitatory transmitters, is more sensitive than usual. The exact genetic problems are not known as yet, but many will likely be found in the next years or decades. The seizure types seen in these syndromes, however, are varied. Juvenile myoclonic epilepsy (JME), for example, has many distinct seizure types—absence staring spells, sudden myoclonic jerks, and generalized tonic-clonic convulsions. Yet presumably, this syndrome arises from a single mutation. Somehow, with a single tiny change in an ion channel, a receptor, or an enzyme in the brain, varied random irritabilities become manifest as seizures. Even in these people, the genetics are not simple. At most, the chances of a child born of a mother or father with JME having epilepsy are about 10 percent. This leads us to suspect that probably not just one gene is involved; a combination of other factors (genetic or environmental) must ultimately result in epilepsy. This complexity may be one reason why it has been so difficult to determine specific causes of genetic epilepsy.

If evolution dictates that over time the strongest and smartest people survive, why would such changes persist? It could be that these genetic differences actually confer an advantage. These slightly overactive brains may function more quickly, with better memory and more creativity than others. Maybe a person with one such gene from one parent has an advantage, while in a child with two copies the brain becomes overactive and seizure-prone. Such a situation is seen in sickle-cell anemia, where one gene confers an advantage in combating malaria, but two result in the disease.

It may also be that genetics plays a role in later susceptibility to epilepsy from other causes. Perhaps your genetic makeup makes your neurons a little irritable, but in a way that no amount of daily stress or sleep deprivation would ever cause a seizure. But years after birth, you have an automobile accident where your brain gets knocked about. In most people, this, too, would not be enough to set up conditions for epilepsy, but the combination of neuronal sensitivity and a car accident yields a partial epilepsy from the injured area.

Congenital Causes of Epilepsy: Problems with Brain Development

The brain undergoes a complicated process during development, whereby neurons are born near the ventricles (fluid-filled areas in the center of the brain) and must travel up through previously developed neurons to the surface of the brain. They wind up carefully placed in six distinct layers in the cerebral cortex. Some of the more primitive parts of the brain (such as the hippocampus, a region in the temporal lobe) have fewer layers, but throughout the brain the placement of neurons is normally quite precise. If this system doesn't work quite right, neurons could develop ectopically—outside of areas where they belong. If this happens, the wiring or circuitry of the brain is not right, and an abnormal, epileptic circuit can result.

Another congenital problem is birth injury. Being born is difficult, and a difficult delivery can leave parts of the brain temporarily starved for oxygen. Perhaps the death of even a few neurons in the wrong place can set up an abnormal, hyperactive circuit in the brain. In the same way, strokes in an infant (due to birth trauma, bleeding abnormalities, or vascular problems) can destroy a part of the brain's normal circuitry, causing seizures, too. Also, a whole array of metabolic problems can affect brain development, such as phenylketonuria (for which there are warnings on soft drink cans, because aspartame is toxic to these children's brains). These people often have decreased intelligence if the problem cannot be corrected, and many also have seizures.

Infections

An infection of the brain (encephalitis) can occur at any age and can usually be treated without any resulting long-term problems. Sometimes, though, encephalitis can result in the later development of epilepsy. This sort of infection is relatively uncommon, but it usually requires hospitalization. Encephalitis can be due to a virus, bacteria, or (rarely) fungus. It is treated with drugs to eliminate the infection (although some viruses do not need treatment). Seizures can occur during the infection itself; in this case, the person does not have epilepsy, but "symptomatic seizures"—seizures occurring because of a known injury to the brain. Such seizures stop once the irritation (the infection) is stopped. A more common infection is meningitis: infection of the coverings of the brain. Strictly speaking, this type of infection does not directly involve the brain and therefore should not be a cause of epilepsy. We do know, however, that it can cause epilepsy. Most likely, such an infection spreads to limited areas of the brain (a meningoencephalitis).

Meningitis or encephalitis is probably most likely to result in epilepsy when it occurs at an early age, and this is one of the more common causes of epilepsy in children. Some scarring of the brain probably happens in these people, such that an overactive neuronal circuit remains even after the infection is gone. Even so, only a small proportion of children with infections will later develop seizures.

Trauma

Children or adults can have a traumatic injury to the brain, either from accidents or, unfortunately, sometimes from physical abuse. The more serious the injury, the more likely that the person could later develop epilepsy. If an injury causes loss of consciousness, concussion, or bleeding in the brain, it is more likely to be associated with epilepsy. Automobile accidents are probably the leading cause of head injuries in most age groups. In the elderly, the shrinking brain floats more loosely in the skull, and the aging body stumbles more easily, so sometimes a serious head injury can happen with a seemingly minor fall.

With any closed-head injury (as opposed to an open injury, such as a gunshot wound), the most likely parts to be injured are the "bumpers" of the brain. The frontal pole—all the way forward in the frontal lobe, above the eyes—can smash directly against the frontal bone. The edges of the temporal lobes also rest directly against bone, as do the occipital lobes at the back. Trauma can directly crush neurons and often leads to localized bleeding. Both can cause seizures.

Seizures can occur acutely, at the time of the injury itself, and again these are symptomatic seizures as opposed to epilepsy. In general, the chances of developing epilepsy are not increased unless there is loss of consciousness or memory for at least thirty minutes. Most children or adults who will develop epilepsy do so within the first six months after the injury, but for others seizures may begin up to several years later. As with an infection, in these people there must be a formation of an overactive circuit of neurons in their healing, such that seizures persist long after the injury itself has healed.

Cerebrovascular Disease

The most common cause of epilepsy in adults is cerebrovascular disease (problems with blood vessels in the brain, resulting in stroke), as even a small stroke can form an irritated area that can later form an epileptic focus. Theoretically, small areas of the brain could be starved for oxygen without actually dying; a stroke has not occurred, but the irritation begins a process resulting in epilepsy.

As many as 10 percent of people with stroke have a seizure at the time of the stroke itself. This does not mean the person will continue to have seizures, but it increases the chances. Other people can develop seizures after recovering from the stroke. This probably occurs as the brain rebuilds injured neurons and circuits, creating an overactive circuit. Most epilepsy due to cerebrovascular disease is easily treated with medications.

Tumors

Brain tumors occur at any age, but are more common in older individuals. Tumors can disrupt brain pathways simply by pushing them around during tumor growth, or by infiltrating and interrupting circuits. Tumors can also be directly irritating to the brain.

Treating the tumor with surgery, chemotherapy, or radiation therapy nearly always improves the seizures, although they can persist, and radiation treatments can cause later disruption of surrounding brain with the potential for seizure recurrence. Also, some people have no seizures with the tumor present, but either the disruption of pathways caused by the tumor or the treatment eventually results in the formation of a seizure focus.

Degenerative Diseases

In older people, diseases such as Alzheimer's cause death of neurons and changes to the structure of the brain. This can result in epilepsy. Seizures are relatively common in Alzheimer's, where they may be relatively subtle and even overlooked in a person who is generally confused. Other degenerative diseases, like Parkinson's, do not seem to commonly cause epilepsy, probably because the region of the brain affected is not susceptible to seizures.

The Role of Alcohol and Drugs

Many people wonder whether epilepsy can be caused by alcohol or drug abuse. Both can certainly cause symptomatic seizures. With alcohol, intoxication (no matter how severe) does not typically cause seizures. People who become alcohol dependent, however, are at risk of alcohol withdrawal seizures. When the body becomes used to a substance like alcohol, it actually changes to accommodate the constant presence of the drug. This is why people who are alcohol dependent feel the need for an eye-opener in the morning: their body feels abnormal without the alcohol. If someone who has been drinking steadily for weeks or longer suddenly stops, alcohol withdrawal begins. Usually shaking and gooseflesh occur (hence the term *cold turkey*). This can be followed by seizures (generalized tonic-clonic). In the final phase, called delirium tremens, seizures resolve but the person can hallucinate. These seizures are not epilepsy, but caused by withdrawal. Other sedatives, particularly barbiturates and benzodiazepines (both of these are commonly used as sleeping pills or antianxiety agents), can similarly cause seizures if used for a long time and suddenly stopped.

Certain other drugs can cause seizures during intoxication. The most notorious are stimulants, such as cocaine and amphetamines. Again, the seizures are symptomatic and are of the generalized tonic-clonic type. Tobacco, caffeine, and marijuana are not known to cause seizures.

The above descriptions are of symptomatic seizures associated with drug use. All too often, though, someone who seriously abuses alcohol or drugs can have other injuries, especially head trauma due to violence or falling. In these cases, drug use can at least indirectly result in epilepsy because of a brain injury. Even without trauma, repeated seizures due to intoxication or withdrawal may ultimately result in epilepsy, although this has never been proven.

Epilepsy of Unknown Cause

Perhaps the most fascinating (and frustrating) situation is also the most common: epilepsy for which no cause can be found. This is sometimes called cryptogenic epilepsy, which simply means that there has probably been a brain injury but it cannot be found. The brains of people with cryptogenic epilepsy look completely normal, as far as modern imaging can show. I suspect that in most of these people several problems have combined to create an overactive area. They may have been born with a bit more of an excitatory neurotransmitter, glutamate, in their brains than others, or a little less of an inhibitory one, GABA. Maybe a tiny injury occurs—a drop in blood flow during birth, a knock on the head—too small to cause any discernible structural damage, but enough to irritate a few neurons. Their extra firing starts to lay down a pathway by long-term potentiation—a normal property of the brain that is probably part of the way we learn: once a pathway is traced, the neurons in it become more strongly linked. Repeat a phone number or associate a name with a face or tickle the irritated neurons, and the pathway (normal or abnormal) becomes stronger, eventually leading to seizures if it's an abnormal pathway. Over time, seizures can become spontaneous and more difficult to control. This is what happened to Marissa after the bleeding of her cavernous angioma, but probably a vast array of problems can start the same process.

There are problems with this theory. The main one is the lack of proof that repeated seizures make epilepsy more difficult to control, although many doctors who treat epilepsy suspect this is the case. Because of this, aggressive and complete control early on—even if this means surgery—may be indicated to prevent increasingly uncontrolled epilepsy.

So what happened with Marissa? It is not clear when her brain became affected, but it was almost certainly at a young age. As she grew up, her left temporal lobe had a scar on it, a small area that was not functioning and began to have seizures. Because this area was busy with its own agenda, it did not develop as fully as it would otherwise have. That is likely the reason she became so creative—her right, more musical brain blossomed in the absence of normal activity on the left. Thus, her epilepsy had a huge impact on the shaping of her personality. But all of us have strengths and weaknesses, areas that interest us and others that we consider horribly boring. Does a similar, perhaps less pronounced process happen in all of us? When we can't get a song out of our heads all morning, how far is this from the early repetitive seizures Marissa had?

Marissa is a wonderful example of epileptic seizures that seem more like an extreme of normal than abnormal. Her small seizures were not a problem; sometimes they were even amusing to her. But the larger, tonic-clonic seizures she eventually had showed that all were in fact part of her disease. Once her brain lesion was discovered, it was overwhelmingly suspected to be the source of her seizures. She was admitted for seizure recording, and at the same time phenytoin (which had always made her a little drowsy) was stopped. The recordings showed exactly what we suspected: the seizures began in the precise region that showed the cavernous malformation. She was started on another drug, lamotrigine, which did not cause her any drowsiness. She also underwent surgery to have the abnormal part of her temporal lobe removed. After surgery, she still had a few vague auras (tiny seizures), although it was harder for her to tell if she was simply

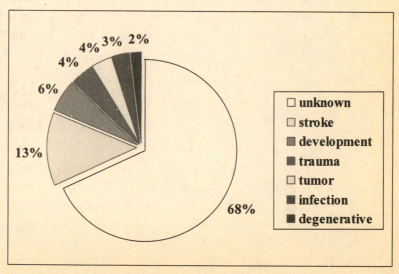

FIGURE 3-1: Causes of Epilepsy

thinking about a phrase repeating or whether it was involuntarily happening. I suspect that other nearby pathways had been trained over the years to do the same thing. Ultimately, though, with the main irritant and the strongest pathway gone, even the tiny seizures faded and ultimately stopped completely like a forgotten memory.

4

Children Are Different

Children are not small adults. Medically, neurologically, and socially they differ from grown-ups in important ways with regard to illness in general and epilepsy in particular. Their brains are developing and changing rapidly and in many ways seem to work differently from adult brains. For instance, neurotransmitters that excite other neurons in an infant's brain cause neuronal inhibition in an adult. Young brains have a much greater capacity to recover and adapt than adult brains, so that certain injuries or surgeries that would cause a permanent deficit in an adult can result in only a temporary problem in a child. Epileptic discharges on an EEG can have patterns in young children that would never occur in adults. Children's lifestyles are also different from adults': children do not drive, and some of their most important interactions are with peers who are not likely to understand their disease. These issues are exemplified in Josh, a ten-year-old child who had developed epilepsy at age nine. In his seizures, he curled his hand and was unresponsive for a few seconds. This almost always happened at school, and his teachers and peers quickly grew to recognize it. Children can be cruel about any child who is different; in some cases, epilepsy can result in

49

prejudice and ostracism. Fortunately, Josh had very involved parents and a school nurse who educated his peers, and he adjusted well to a young life with seizures. Now, on a low dose of medication, he has no seizures. Seizures in this age group usually improve or resolve with time, and should Josh remain seizure-free, medication will likely be withdrawn before he gets close to driving age.

This chapter discusses the seizures occurring most commonly or exclusively in children, and issues of treatment and lifestyle unique to children.

Seizures in Infants

Febrile Seizures

Fever is one of the most common causes of seizures not only in infants but in general. Up to 5 percent—one in twenty—of infants will experience a febrile seizure. These seizures take place during an illness with a high fever (often 104°F or even higher), although they can happen at lower temperatures. The baby will typically experience a generalized tonic-clonic convulsion, with sudden stiffening of the arms and legs, followed by jerky movements of the limbs. She may be sleepy and difficult to arouse afterward.

Although common and nearly always benign, these seizures need to be evaluated by a pediatrician or a child neurologist. Febrile seizures, particularly if they are happening for the first time, can reflect a much more serious illness such as meningitis or encephalitis (infections of the coverings of the brain or the brain itself, respectively). If there is an underlying injury to the brain, such as a birth trauma or tumor, seizures can first manifest themselves during a fever. On the whole, though, febrile seizures usually carry no long-term consequences. They probably do not injure the brain unless prolonged (an hour or more). Febrile seizures do tend to run in families, and if a parent or a sibling has had them, it is even more likely that they do not portend anything serious. As these are considered symptomatic seizures, they are not a form of epilepsy.

Febrile seizures are almost never treated, even if recurrent. Sometimes, medication such as diazepam (Valium) is given when fever occurs to prevent the seizure, but as the seizure may not occur (and carries no long-term problems even if it does), it generally does not make sense to do this. In rare circumstances (such as frequent or severe febrile seizures) a doctor may recommend giving an anticonvulsant. This should be short-term and stopped when the risk of seizures dissipates (usually by about age five).

Infantile Spasms

Infantile spasms typically begin between the age of six months and two years. The baby will suddenly stiffen, with arms extended and eyes wide open. After a few seconds, he will relax. This spasm then repeats every few seconds, and the cluster can last many minutes. The baby often cries after each spasm.

Children with infantile spasms need to be evaluated immediately by a neurologist. There may be abnormal brain activity even when the child is not having spasms. Both the spasms and the potentially abnormal brain activity can interfere with the child's development at a critical time and need to be treated. The anticonvulsants used most often are ACTH, valproic acid, topiramate, and benzodiazepines. ACTH is actually a steroid hormone and can have numerous toxic effects (particularly hypertension and thickening of the heart wall). ACTH is usually started in the hospital, and the child taking it must be closely monitored for these problems.

Another drug, vigabatrin, may be effective (particularly if the child has a condition called tuberous sclerosis, where numerous benign tumors are present in the brain starting at birth). Vigabatrin is not available in the United States. It is sometimes recommended, however, and Americans can obtain it from other countries (usually Canada or Caribbean ones) with a prescription written by an American physician. Although there has been talk of restricting the import of prescription drugs from other countries, the FDA now allows this for personal use. Shipments are occasionally delayed in customs, however, so people who obtain drugs from overseas should use caution to always have at least two weeks' worth in reserve.

Seizure Syndromes Unique to Childhood

Absence (Petit Mal)

Absence seizures are subtle and can often be missed for months or even years. Simple absence seizures consist of only a brief interruption (usually one or two seconds) in activity. If in the middle of a sentence, the child may pause, then finish the sentence. If playing a video game, her activity may stop, then immediately resume. There are no other changes, although with a prolonged absence (lasting five or more seconds) the child could smack her lips, flutter her eyes, or drool slightly. After the seizure, she is completely normal immediately. The child may not even realize that a pause has occurred, although older children typically realize that they have missed something. Teachers may note the staring spells before parents do.

Absence seizures should not be confused with daydreaming, although without careful examination the distinction can be difficult. The main difference is that daydreaming usually lasts longer, and the child can be aware during the episode. Most important, a daydreaming child can be alerted during the episode, but a child having an absence seizure cannot. Again, given the brevity of the episodes, it is sometimes impossible to make the distinction without evaluation.

The only way to definitively diagnose absence seizures is with an EEG. This will show a typical pattern, where brain cells fire in synchrony at a specific rate (around three times per second). The pattern can be seen between seizures, but the most reliable diagnosis is

made by recording an actual staring spell either occurring spontaneously or by having the child hyperventilate, which often induces
absence seizures.

Absence seizures are usually stopped with medication. If they are
mild, sometimes the doctor will not recommend treatment. In most
cases absence seizures stop spontaneously during adolescence, and
medication will no longer be required, although in a few cases they
persist into adulthood and treatment needs to be continued.

Benign Rolandic Epilepsy (BREC)

Benign rolandic epilepsy of childhood (also called by the
acronym BREC, or childhood epilepsy with centrotemporal spikes)
usually begins at age six to eight years. Seizures occur exclusively or
almost exclusively in sleep. The child usually awakens with a sensation of tingling on one side of the face, or with one side twitching.
There is frequently a lot of drooling, and the child may not be able
to speak. The episodes can be frightening to both child and parent.
They last a minute or so, then stop just as suddenly as they start.
Rarely, similar seizures occur with the child awake. Generalized
convulsions can also occur; these usually start similarly but
progress into a grand mal seizure.

This epilepsy is called benign because it always stops in adolescence. Because the seizures almost always happen in sleep, they are
usually not disruptive unless the child has a sleepover or is at camp.
Diagnosis is usually straightforward from the history, but can be
confirmed with an EEG because of a characteristic electrical pattern
that occurs particularly during sleep. An imaging study of the brain
(an MRI) is usually not necessary, but could be recommended to ensure that the brain is structurally normal if there is any question in
the diagnosis. If the seizures are frequent or disruptive, they are easily treated with a low dose of an anticonvulsant, which can be
stopped in adolescence.

Photosensitive Epilepsy

Flashing lights can be a trigger for epileptic seizures, although this is less common than many believe. This phenomenon became particularly well known after several hundred Japanese children had seizures induced by rapidly flashing lights during an episode of *Pokémon,* aired on television on December 16, 1997. In this case, millions were watching the program and the producers inadvertently used a flash frequency that is particularly liable to cause seizures in susceptible children (about twelve hertz, or twelve times per second). The condition is rare, but because such a huge number of children were watching, several hundred were affected and a few required emergency room visits.

The vast majority of children with epilepsy are *not* photosensitive. In those who are, the frequencies they are exposed to via television, computers, and video games are not usually problematic, although some caution is advisable. Susceptible children can occasionally experience seizures from traveling rapidly down a tree-lined street, with resulting repetitive bursts of sunlight between the leaves. Rarely, children can even learn to cause seizures by moving their hands in front of their eyes to cause flashing.

Photosensitivity can be shown on an EEG. As part of a routine examination, flashing lights of various frequencies are displayed to determine whether they induce epileptic changes in brain waves. Therefore, it is relatively easy to determine whether a child is at risk for seizures induced by flashing lights.

Other Seizure Types

Children can get all of the seizure types described in chapters 1 and 2, including all types of partial seizures. Young children may have difficulty describing the strange feelings they are experiencing, making it difficult to determine the onset site of a focal seizure. They may be able to describe visual changes, particularly if asked in a nonthreatening way; a four-year-old patient of mine had no trouble talking about a "black butterfly" that she saw fluttering in her left field of vision before her seizures.

Treatment Issues in Children

Even more than in adults, it is critical to correctly diagnose the type of seizures occurring in a child. Staring, for instance, could be a symptom of either absence or complex partial seizures, and these are treated very differently.

More in children than adults, it is sometimes reasonable not to give medication even after seizures have been diagnosed. As children are less likely to be at high risk for injury (they are not driving or operating heavy machinery), it might make sense to tolerate a rare, relatively minor seizure rather than expose the child to daily medication with its potential for side effects. Also, many seizure syndromes in children resolve spontaneously with no long-term consequences, particularly absence and benign rolandic seizures. Again, simple caution may be advisable as opposed to the (admittedly small) risks of medication.

Most children with epilepsy, however, do require medication. Many physicians and parents worry about the potential influences of epilepsy medicines on critical learning processes that take place during childhood. If a child falls behind in school because he is always sleepy or inattentive, it may be difficult for him to catch up. In general, the influences of epilepsy medications on learning are theoretical rather than known. There may be small differences in a child taking medicine, but these, if present at all, could only be seen with careful testing. Nonetheless, certain guidelines are helpful if a child requires medicine. First, it should be given at the lowest dose required to control seizures. An infrequent, mild seizure that "breaks through" the medication may actually be less of a problem than daily high doses of sedating drugs. Second, if at all possible, one medication should be used rather than two, although in some cases two or even more drugs are needed to control the seizures. The potential for side effects is always less with one drug. Finally, sedating drugs should be avoided in most older children, particularly phenobarbital. This drug is used widely in infants, but it may impair performance in school-age children because of its sedative effects.

Behavior is also a concern in children whenever drugs are used that affect the brain. Even with no effect on learning per se, a drug that affects a child's ability to sit still in class, or one that makes her irritable and difficult to control, may be much more disruptive and harmful than a rare seizure. Again, most children do not develop personality changes or irritability when taking drugs for epilepsy, but these effects are possible. Gabapentin, for example, is an extremely safe drug and therefore is often used in children, but a small percentage may develop behavioral problems when taking it. Even though phenobarbital is a sedative, it can cause hyperactivity in some children. The best approach is to be aware of the possibility of behavioral effects, and to discuss with the prescribing doctor if a medication-based behavioral change is suspected. Parents are in a much better position to notice such changes than a physician, who sees a child periodically and for only a short time.

Some drugs have side effects in children that do not occur in adults. A rash can occur with any drug, but children seem more susceptible to serious rashes from lamotrigine. In very young children (less than age two), valproic acid should be avoided because of a higher risk of liver failure. Topiramate in adults sometimes causes difficulty in thinking clearly, but this seems to occur less often in children. Felbamate is not used often in adults because of more frequent toxic reactions affecting blood and the liver, but some think that this drug is actually safer in children.

As in adults, anticonvulsant drugs don't always work in children. Surgery can be an option whenever seizures persist and adversely affect a child's life. Many epilepsy specialists actually think that children should be referred for extensive evaluation sooner than adults, as years lost in trials of drugs may lead to a lifetime of emotional dependency and limited abilities, whereas early treatment of a child with epilepsy will allow her to develop normally, socially and intellectually. Surgery always carries some risk of adverse effects, but children's brains are also the most capable of recovery in their younger years, so early surgery can increase the chances of completely normal function later in life. The evaluation for surgery in children is similar to that for adults and is discussed in chapter 9.

Social Issues in Children with Epilepsy

Sometimes doctors (and parents) can become so caught up in any child's disease, including epilepsy, that they can overlook the effect the disease and the treatment may have on other aspects of his life. Children may not talk about the embarrassment they feel when seizures are seen by classmates, or when they must go to the school nurse daily to get medication. Teachers or coaches who are unfamiliar with the disease may place too many restrictions on the child's activities, making him feel different from others. Older children may be troubled by what (or what not) to tell peers about their disease. Once dating begins, worry about what to say about seizures, and when, becomes particularly troubling and comes in addition to the struggle of self-development that all teenagers face.

Parents should be careful not to be overly protective. While more severe seizures carry risks, and certain safety precautions may be required (such as not swimming alone and in older children not driving), children should be allowed to live as normal a life as is medically reasonable. If a child is constantly watched and restricted, he may develop a tendency toward dependence and self-doubt that can persist even after seizures are stopped.

The most important thing that parents and doctors can do is to make sure that the child feels comfortable talking about every aspect of the disease. Asking direct questions about his feelings regarding the disease, such as "Do you ever have trouble with other kids not understanding what happens?" or "Does taking medicine every day bother you?" is a start. Educating teachers and peers through pamphlets or educational tapes can make a big difference in their acceptance of epilepsy and of the child. Older children may particularly benefit from talking to others with similar problems, perhaps through the Epilepsy Foundation. As in all aspects of their lives, adolescents value the opinions and experiences of peers and will likely have an easier time accepting medication and minor restrictions on activities if they hear about them from a peer rather than only from their doctor and parents.

5

Seizures in Older Adults

A popular misconception remains that seizures always begin in the young. Actually, epilepsy begins most frequently at two times of life: in children, and in persons over sixty. Because of the misperception, however, many older people remain misdiagnosed or undiagnosed. Seizures are commonly mistaken for memory lapses, for syncope (fainting because of a sudden drop in blood pressure), or strokes.

Older people who have epilepsy also need slightly different treatment. In general they need lower doses of medication for a variety of reasons: slower metabolism, decreased kidney function, greater sensitivity to the adverse effects of drugs. The causes of seizures in older people are also somewhat different, so that onset of seizures in an older person requires different testing to look specifically for the reasons most common in this age group.

This chapter deals specifically with issues related to older persons. There is no single age when these become important; there is some relevance to persons over fifty, and the issues become more important at older ages.

Seizure Types and Causes of Epilepsy in Older People

The chance of getting epilepsy rises as people age; over about age sixty-five, onset begins to increase fairly dramatically. The seizure types in this age group are quite different from those of childhood and, to some extent of younger adults. In older people, epilepsy is virtually without exception of the partial type. Remember that generalized seizures usually have some sort of genetic predisposition and will therefore begin much earlier in life if present. After adolescence, generalized seizures (such as absence or myoclonic) almost never begin (although they can continue).

The differences in seizure types become clearer when considering the causes of new epilepsy in older people. Almost all cases are due to an injury to part of the brain: a stroke, a tumor, or a traumatic injury from a motor vehicle accident or a fall. Epilepsy can also occur from other diseases that injure the brain in other ways, such as Alzheimer's disease.

Sometimes the cause is easily seen, as in someone who had a stroke and a month or so later begins to have seizures beginning in the same part of the brain. Much more often, however, there is no obvious cause (as is the case with all age groups). In these cases it makes sense to look more closely into the possibility of an unrecognized stroke or tumor. People should be evaluated for stroke risk factors, such as high blood pressure, heart disease, diabetes, elevated cholesterol, and smoking. As with most younger persons, a detailed picture of the brain (an MRI) is required. Even if the MRI is normal, I sometimes recommend studies of the arteries that feed blood to the brain. Even without structural damage seen on the MRI, there may still be severe decreases in blood flow, causing parts of the brain to be deprived of nutrients or oxygen. This is irritating to the brain and could result in seizures. Perhaps more im-

portant, correction of the problem could prevent a mor᠎
ing stroke.

Consideration of Drug Choices

Older people have several differences from the young that can affect the choice of drugs. As our bodies age, the kidneys tend to slow down. In most this is not a problem, but for drugs that are eliminated from the body primarily or exclusively by the kidneys (such as gabapentin), the same dose in an older person will result in higher levels in the body. Most drugs are at least partly eliminated by the liver, which also works a bit slower with advanced age. With nearly all drugs, then, the same dose in an older person will result in higher amounts in the body than in a younger person.

Sensitivity to drugs in older persons comes from several causes, including higher blood levels as described above. Independent of the actual level of the drug in the body, however, older people are more likely to become dizzy, tired, or confused, or to have double vision at levels that a younger person would tolerate without a problem. For this reason drugs such as gabapentin or lamotrigine, which are less likely to cause these effects, may be preferable. Phenobarbital may be problematic because of its sedative properties, to which older people may be more sensitive. Phenytoin is less likely to cause sedation, but in many older persons phenytoin is not the best treatment choice. Its level in the body is extremely sensitive to many factors, including other drugs taken, absorption (which can be affected by illnesses affecting the digestive system), and liver function. For this reason, an unexpected increase in the blood level of phenytoin can occur, with resulting dizziness, double vision, and sedation. In fact, toxic levels of phenytoin are a common cause for hospital admission, particularly in older persons. One large study of older people with epilepsy showed that both lamotrigine and gabapentin were better tolerated than carbamazepine, another commonly used drug, and were at least as effective, so these may be particularly good choices.

For reasons that are not well understood, older people may also have their seizures fully controlled at doses and blood levels that would be considered "subtherapeutic," or below the minimum effective amount, in younger people. Recommending a lower dose, when appropriate, can further help to avoid toxic symptoms.

Finally, older people are more likely to have other medical conditions along with epilepsy. This is mainly important because the additional medications they may be taking could interact with the epilepsy drug. Many medications can change the concentration of the epilepsy drug, and vice versa, resulting in potentially ineffective or toxic amounts. As we'll discuss more extensively in chapter 8, many epilepsy drugs alter liver enzymes that metabolize drugs needed for diabetes, blood pressure, infection, and other problems. Older persons commonly take six or more drugs, making potential interactions among the drugs complicated. The use of one epilepsy drug with few or no interactions with other drugs is therefore often preferable. Gabapentin and levetiracetam have no known interactions, making them good choices for many people who take a lot of other drugs. Several others, including lamotrigine, topiramate, and zonisamide, have no known effects on other drugs and may be good choices in some people; however, some drugs can affect their levels and effectiveness.

Lifestyle Issues

Besides being more sensitive to toxic effects of medications, older persons may be more susceptible to adverse effects from seizures. Whereas a healthy young person may fully recover from a generalized tonic-clonic seizure within hours or even less, older people can take several days to feel themselves again. As we age, we lose calcium, making bones easier to fracture in seizure-related falls. Bone loss generally occurs faster in women, but also occurs in men. Bone demineralization can be accelerated by some anticonvulsant drugs, including phenobarbital, phenytoin, valproic acid, and carbamazepine. It is therefore even more important to err on the side of caution in completely controlling seizures.

Older people are more likely to have memory problems than younger ones. Some of this is due to diseases such as Alzheimer's, which occurs in over 40 percent of individuals over the age of eighty-five. In all people, however, memory changes as we age. It becomes more difficult to remember things in the short term: where you left the keys or the name of the person you met half an hour ago. Many people also experience increasing difficulty coming up with words—the "tip of the tongue" phenomenon in which you know exactly what you want to say, but just can't seem to get the word out. Often this becomes frightening, as a possible sign of a serious problem. If in doubt, it is a good idea to ask a doctor about these symptoms. Other tests, particularly neuropsychological testing, may be necessary to determine whether there is, in fact, a problem. Fortunately, in the majority of cases memory "difficulties" turn out to be just the benign effects of aging. Reassurance, and sometimes counseling in techniques to remember things better, are all that are needed.

Because benign memory problems are so common, however, it often makes sense to make medication as simple as possible for this group. Medications such as zonisamide, which may be taken once a day, may be preferable. Lamotrigine can also sometimes be taken once a day. Phenytoin and phenobarbital can be dosed once a day, but as discussed, their side effects often make them less desirable choices. Their sedating effects can contribute to sluggishness and to slower recall and more perceived problems with memory. The slower metabolism seen in older people makes twice-daily dosing reasonable for most other drugs, which is much easier than three or four times a day.

Other Treatments

Older people may also benefit from treatments other than medication, including surgery. As with other age groups, if seizures are not fully controlled after adequate trials of two medications, video-EEG recording should be considered to confirm epilepsy and see if surgery is an option. In particular, spells in older people that do not respond to seizure treatment may turn out to be from cardiac causes (such as arrhythmias of the heart) that cause sudden loss of blood flow to the brain and loss of consciousness. Transient ischemic attacks (TIAs, or "ministrokes") also occur in this age group and can be confused with seizures. Psychological problems can occur as in younger persons, and certain sleep disorders become more common. For example, strange or violent behavior that occurs only during sleep may be REM behavior disorder, a relatively rare condition that can also be confused with epilepsy. If video-EEG monitoring confirms epilepsy, surgical treatment may be considered. Although the risk of any kind of surgery is slightly greater in older people, seizures carry real risks as well, including injury or even death. As in younger people, this option should be discussed carefully with a neurologist who is experienced in surgical treatment of epilepsy. Most epilepsy centers will consider surgery in people well into their sixties, and sometimes beyond. In older persons whose seizures persist, careful consideration needs to be given to driving and living alone. Overall, however, most older people's epilepsy can be fully controlled on low doses of medication.

6

When Seizures Are Not Epilepsy: Other Conditions That Look Like Epilepsy

Because of the relatively few reactions the brain can have to injury, a variety of other conditions can look like epilepsy and make diagnosis a challenge. As the doctor rarely sees an actual seizure, but rather relies on secondhand descriptions, the diagnosis can initially be epilepsy, but with time, treatment, and further episodes this may change. A correct diagnosis is obviously critical both in counseling the person and in finding the best treatment for the problem.

Since the age of fifteen, Ruby had had episodes in which she saw flashing lights on one side. These would grow in intensity for about fifteen minutes, then a tremendous headache would start. She usually became nauseated and needed to cover her eyes because any light at all made the nausea worse. Often, she became unable to speak during this time. Once or twice, the episode progressed into loss of consciousness and a generalized convulsion.

Ruby had seizures, but she did not have epilepsy. She had severe migraine headaches that can rarely (as in her case) progress into seizures. She did not, therefore, need treatment for epilepsy, but rather for migraine. Once her migraines were adequately controlled, she no longer had these severe episodes resulting in seizure.

Many other conditions can be confused with epilepsy. Most are also neurological (besides migraine, strokes, and movement disorders), but others may be cardiac or psychological problems. In this chapter we'll explore the major conditions that can be confused with epilepsy. Remember again that "symptomatic" seizures (a seizure because of an infection of the brain or from drug use, for example) are not epilepsy either; this will be discussed again in chapter 7, on the diagnostic process.

Neurological Conditions

Migraine

Most people think of migraine as a bad headache. This is true in the majority of cases: the major symptom is a throbbing headache, usually on only one side of the head, accompanied by nausea, vomiting, and sensitivity to loud noises or bright lights. This is known as a common migraine and would never likely be confused with epilepsy (however, many people with epilepsy have migraine—see chapter 14). Other people have a warning, an "aura" before a migraine headache begins. This type, "classic migraine," can be confused with epilepsy because migrainous and epileptic auras can be similar. The most common migraine aura is a sensation of flashing lights surrounding an area of temporary blindness (known as a scotoma). Symptoms of visual change can occur in seizures that begin in the occipital lobe and can actually be quite similar to those experienced during a migraine aura, but blindness is unusual during a seizure. Another, more unusual symptom of migraine includes weakness on one side of the body. This is called hemiplegic migraine and is more often confused with stroke than epilepsy, because seizures are rarely associated with loss of movement (they typically cause *increased* movement, such as clonic jerking). Migraine, like epilepsy, can also include difficulty speaking and even loss of consciousness.

In these unusual cases, the distinction between migraine and epilepsy can often be made by the duration of symptoms. It is unusual for a seizure to last more than a minute or two, but migraine auras usually last fifteen minutes or more. The presence of headache is helpful in the diagnosis, particularly when these are typical migraine headaches, but migraines can occur without headache, and headache can accompany seizures. In the case of Ruby, a classic migraine led to an epileptic seizure. Probably this occurs because of ischemia, or decreased blood flow, during the migraine, which causes irritation to the brain, resulting in seizure. This combination is unusual, but the distinction is important because (as in Ruby's case) the correct treatment is treatment of the migraine, not the seizure.

Stroke/TIA

Stroke and transient ischemic attacks (TIAs, sometimes called ministrokes) are caused by permanent or temporary interruptions in blood flow to a part of the brain. This can result from diseases such as atherosclerosis, causing increased narrowing of blood vessels with ultimate occlusion. TIAs and strokes can also be caused by anything in the bloodstream that can block off an artery; most commonly, this is a blood clot from another part of the body. As with seizures, the location of a stroke or TIA will determine the symptoms. If areas controlling movement are involved, the person may become paralyzed on one side of the body. If visual areas are affected, there may be partial or complete blindness. With TIAs, the interruption in blood flow is temporary, and the symptom resolves as blood flow resumes. Once it is over, the person is completely normal again. With a stroke, blood flow is interrupted permanently or long enough that neurons in part of the brain die due to lack of oxygen. Recovery occurs, but it is slower and may not be complete.

The main difference between stroke/TIA and seizures is the nature of the associated symptoms. Seizures tend to produce *positive* symptoms—the brain has increased activity, so the person sees something (hallucinations) or feels something (déjà vu, tingling) or moves abnormally (clonic jerking). Strokes usually have *negative* symptoms—the person can't move, can't see, can't feel. There are exceptions to this: inability to speak, for instance, occurs in both.

Another major difference between stroke and seizure is in the duration of symptoms. While a seizure typically lasts one or two minutes, a TIA usually lasts at least twenty minutes, and a stroke much longer.

The distinction between strokes/TIA and seizures can be complicated in a person who has both (see chapter 3). It is probably not surprising that a relatively high percentage of strokes (about 10 percent) cause enough irritation of the brain to precipitate a seizure as they occur. In such cases, seizures can persist because of the damage done by the stroke, and the person might need treatment for both conditions. When only one condition is occurring, a physician can usually easily distinguish between the two by taking a careful history.

Heart Conditions

Heart problems can be another reason that the brain temporarily does not get enough blood. An arrhythmia—an abnormal heart rhythm—is one of the more common situations in which a heart condition could be confused with epilepsy. For example, if the heart rhythm changes for ten seconds or so, causing the heart to pump blood ineffectively, the brain will become starved for oxygen.

The person can suddenly feel dizzy or light-headed or (more typically) will abruptly pass out. To add to the confusion, tonic movements (stiffening) of the arms or legs can simultaneously occur, or even a seizure, due to the irritation of the brain. Obviously, this is a life-threatening emergency and needs to be treated, usually with a pacemaker. The seizure itself, in this case, if it occurs, is symptomatic and does not require additional treatment.

Syncope

Syncope, also known as fainting or passing out, is a frequent occurrence. Usually, the person first feels light-headed or dizzy. This may persist for many minutes, and he may try to sit or lie down. Particularly if this is not possible, the person may suddenly fall to the floor, usually awakening within seconds.

Syncope usually happens because of a combination of factors that make it harder for the heart to pump blood to the brain. The amount of blood in the body may be reduced by dehydration, as can occur with profuse sweating on a hot day. When you are standing up, the heart has to work harder against gravity to get blood up to the head than when you are lying down. While you are walking, the muscle contractions in the legs help to push blood back to the heart, but if you are standing still, the blood tends to stay in the legs and away from the heart (and the brain). Finally, the heart sometimes has paradoxical reactions that make it difficult to effectively pump blood: fear or bearing down (as when straining to have a bowel movement) can cause the heart rate to slow. Sometimes one of these conditions is enough to cause syncope, but more often, two or more add up to cause the faint.

Syncope can be confused with seizures particularly when the person doesn't remember anything about the event, and when nobody sees what happened. The person may only know that he passed out; an obvious seizure might have happened, but is not remembered. As with cardiac arrhythmias, you can also have a seizure because of syncope. This typically occurs only in more severe cases, as when a person faints but because he is in a chair or someone is holding him up, he remains upright and the heart has continued difficulty getting blood to the brain (this is why people who faint should always be allowed to lie down, at least temporarily). If the diagnosis is syncope, even if a seizure occurs, treatment may be as simple as remaining well hydrated to avoid future occurrences.

One such person, Dan, was recovering from a bad flu and had not eaten or drunk much in the past week. He was walking through a mall and suddenly felt light-headed. He remembered looking for a place to sit, but didn't know what had happened until he awoke in an ambulance and was told he had had a seizure. Everything was normal when he was examined (see chapter 7) except that he had bitten his tongue. Most likely he was dehydrated from his illness, and prolonged standing resulted in syncope followed by seizure. Although he had a seizure, he did not have epilepsy and was simply told to stay well hydrated if he found himself in this condition again. He has had no further problems.

Sleep Disorders

Seizures frequently occur during sleep. Other problems, though, also happen during sleep, and particularly because the person's description of sensations and events is so important in diagnosing epilepsy, it may be hard to know what has happened if the person was sleeping when it started or slept throughout the episode.

Most people have experienced sudden jerking of arms or legs while falling asleep, called hypnic myoclonus. This may be accompanied by a sensation such as falling, or by a frightening image like a man with a gun. It may appear identical to epileptic myoclonus, but it is completely normal.

Children especially commonly experience sleep terrors: suddenly awakening, often screaming, and possibly confused for several minutes. If the onset is not seen (the rest of the household is probably asleep), it might be thought that a seizure occurred, followed by confusion. Sleep terrors, like hypnic myoclonus, are completely benign and rarely require treatment.

A rare condition called REM behavior disorder can also be confused with seizures. During sleep, the muscles of the body normally become more relaxed. Sleep consists of several stages, which are called REM (rapid eye movement) and non-REM (consisting of stages 1 through 4). During the REM stage, the most vivid dreams occur. REM is also the stage where muscles are normally the most relaxed, and movement does not occur. In REM behavior disorder, the relaxation normally present during REM does not happen. Because of this, the person can "act out" dreams. These can be violent. People have been known to injure themselves or their bed partners, to leave the bedroom or the house while still fast asleep, or even to try jumping out of windows. The confusion and bizarre movements in this disease can result in misdiagnosis as epilepsy.

Other conditions that are sometimes, though rarely, misdiagnosed as seizures include confusional arousals ("sleep drunkenness"), sleepwalking (somnambulism), and sleep talking (somniloquy). Confusional arousals are much more common in children. Usually the person is awakened for some reason (by the telephone, another person) and appears confused, possibly not knowing who or where she is, and slurs her speech. This is another benign condition and occurs simply because of slow awakening from the deepest stages of sleep. Many people will recognize this phenomenon from when the telephone has awakened them from a deep sleep, and the next morning they have no recollection of the conversation. Sleepwalking and sleep talking are also common in children and are usually easy to distinguish from epilepsy. These also occur in non-REM sleep, are usually not remembered by the person, and are benign. All are more common under conditions that deepen sleep, such as after sleep deprivation or with consumption of certain sedatives. Most people do not require treatment for confusional arousals, somnambulism, or somniloquy.

Psychiatric Disorders

The distinction between neurological and psychiatric conditions has become more and more blurred as we learn more about the workings of the brain. We now know, for instance, that depression is associated with a chemical imbalance in the brain that can often be corrected with medication.

Several conditions that are generally termed psychiatric can be confused with epilepsy. Panic disorder and conversion disorder share many characteristics with epilepsy. Rarely, some symptoms of schizophrenia can be confused with epilepsy. In all of these cases, the distinction can usually be made with a careful history, but sometimes recording of the actual abnormal event using video and EEG is required to be sure of the diagnosis.

Panic disorder consists of the sudden onset of intense fear, with no apparent reason for its occurring. It is normal, of course, to become afraid if you are in the jungle and suddenly find yourself face-to-face with a tiger or a lion that seems to be eyeing you hungrily. Your heart will race, you will sweat, and you might feel faint. In panic disorder, these symptoms repeatedly appear out of nowhere. Hyperventilation can occur, even without the person realizing it, resulting in further light-headedness or passing out. Symptoms of sudden, intense fear, light-headedness, and passing out are not unusual for a temporal lobe seizure, and panic attacks could be confused with this.

Conversion disorders are best understood as the conversion of an intolerable thought into a physical symptom. A dramatic example is a person I saw who witnessed his place of business burning to the ground, and who suddenly became paralyzed on his right side as if he had had a stroke. In fact, his was a conversion reaction, which resolved completely once the shock had worn off.

Sometimes, for reasons that are not understood, people can develop a conversion reaction that very much looks like a seizure, even to a well-trained physician. These persons can suddenly lose consciousness, sometimes falling and shaking as during a grand mal seizure. In some cases, this can occur because of horrible experiences in the past, such as physical or sexual abuse, which were so terrible that the person learned to shut down mentally while the abuse occurred. In other people there may be no obvious reason for the reaction. These cases can only be diagnosed after careful evaluation by a neurologist or epilepsy specialist, usually with video-EEG monitoring (discussed in chapter 7). They are sometimes called nonepileptic seizures, because of their similarity to epileptic seizures in description. Although not common in the general population, this condition is actually common among people whose epilepsy has not responded to medication. Nonepileptic seizures comprise about 25 percent of persons referred to epilepsy centers.

It may seem strange to think that there is a psychiatric condition in which the brain is normal, but the person can suddenly have an attack resembling an epileptic seizure seemingly at random. Consider, though, that all people selectively block out part or all of their environment at times. The simplest example of this is driving: most of us have had the experience of driving for many minutes in deep thought or conversation, and once arriving at our destination having no memory of getting there. Perhaps in a similar way, some people learn to suddenly and involuntarily block out the world for minutes at a time. The description of this sounds very much like an epileptic seizure, but treatment in these cases consists not of epilepsy drugs, but psychotherapy to look for possible causes and for other conditions such as depression or post-traumatic stress disorder. Relaxation techniques and training in awareness can also help people to regain control during the episodes. Diagnosis of this condition is particularly difficult as it can occur in people who also have epileptic seizures.

Many descriptions of seizures include hallucinations, usually visual or auditory. Most people associate hallucinations more with psychiatric disease, specifically schizophrenia, than with epilepsy. However, the nature of hallucinations in schizophrenia is different from those of epilepsy, so it is usually quite easy to tell the difference. The auditory hallucinations of schizophrenia are usually a voice or many voices, which can be present for hours or days. They will seem to speak to the person, and they are menacing and unpleasant. They tell the person that he is useless or bad, and they sometimes tell him to hurt himself or other people. The hallucinations of epilepsy, by contrast, are brief. They are stereotyped, meaning that each time they occur they will be precisely the same. Visual hallucinations in schizophrenia are actually rare, but will have the same qualities as the auditory ones.

Variations of Normal Phenomena

In the broadest sense, most people have had episodes that could superficially be confused with seizures: an intense feeling of déjà vu, inability to get a phrase of music out of one's head, automatic driving with no memory of what has transpired. Most people accept these as normal. In rare cases, experiences such as these can be unusually intense or troublesome and may come to the attention of a person's physician. Making the distinction between these events and epilepsy is usually possible with a careful history. There should never be any loss of control with normal phenomena. Still, some people have intense sudden feelings or inspirations that are not that different from those experienced by people with mild forms of epilepsy. Distinguishing between the two may not be important unless the episodes in question are causing problems, but mild seizures (even if not troublesome) have the potential to become more serious and should certainly be evaluated.

Seizures Versus Epilepsy

Finally, epileptic seizures can occur because of a multitude of conditions other than epilepsy. In general these are conditions (such as an infection) that irritate the brain. The diagnosis here is not epilepsy, but infection: this is a critical distinction because once the infection resolves, so will the seizures, and treatment for epilepsy will not be required. Conditions other than epilepsy that can cause seizures are discussed in the following chapter on diagnosing epilepsy.

7

Making the Diagnosis: The Process and the Tests Involved

In terms of diagnosis, one aspect of epilepsy is particularly unusual: the doctor almost never sees the actual problem. When appendicitis occurs, a doctor can examine the tender abdomen and measure the fever and the increased white blood cell count. With Parkinson's disease, the tremor and difficulty moving are readily apparent on examination. But because seizures are usually brief and the person then returns to normal, typically the doctor sees nothing in the emergency room or the clinic except a normal person who may understandably be worried, but who is physically and mentally completely normal and often doesn't even remember what has happened.

The First Step: Description of the Event

The diagnosis of epilepsy is based first and foremost on a description of what happened, which may be firsthand in part (depending on how much the person remembers), or secondhand (if a witness is available). Sometimes, the person has no memory whatsoever of the event and nobody saw what happened. In this case, details about what happened before and after can still be helpful in making a diagnosis. Occasionally changes in the person are visible to a physician shortly after seizures or even long after, but most people appear completely normal by the time they are seen by a physician.

The importance of a complete and accurate description cannot be overemphasized. I once saw a five-year-old boy named Matthew whose parents described violent shaking of his arms and legs, lasting about a minute, which he could not remember. They were worried and awakened him each time, to find him confused and irritable. His physician understandably thought they could be tonic-clonic seizures, but referred Matthew to me for video-EEG recording to be sure. She was suspicious that they were *not* seizures because on careful questioning the episodes always seemed to happen early in the night with no stiffening and no outcry. Once these attacks were seen on the video-EEG, it become clear that they were not seizures at all but benign myoclonus of sleep: an exaggeration of the slight jerking most people experience occasionally when falling asleep.

Most people come to medical attention after they have experienced a strange event that may or may not have been recognized as a seizure. In the case of the most dramatic seizure, a generalized tonic-clonic convulsion, an onlooker may immediately recognize this as a seizure, particularly if he or she has seen one before or if he or she has some medical training. Many aspects of the seizure will be important for the physician to know: What was the first unusual episode experienced by the person? If the person experienced an aura (often considered a warning, but actually a small seizure), this can help in diagnosing the type of seizure and where in the brain it might be starting. Did anything unusual take place shortly beforehand? If a seizure occurs with an illness, drug or alcohol use, or head injury, this can help in determining its possible causes. What happened afterward? Did the person awaken immediately, or was there a period of unconsciousness or confusion?

If more than one seizure occurred, the setting and description of each can help. It is not unusual for a person to experience many small seizures before recognizing that something is wrong.

Ideally a witness can describe the seizure in detail from start to finish. It is always helpful to bring along a witness when seeing a doctor for a suspected seizure even if the person was only present for part of the seizure. Even if the patient is convinced he or she remembers the entire event and appeared awake to onlookers, a brief period is commonly forgotten. Often people experience odd, frightening, or bizarre feelings that may even be embarrassing to relate, but these are also important to share with the doctor. Did a strange, almost psychic feeling come out of nowhere? A strange smell, a tingling in the left arm?

In most cases, a careful history enables the doctor to make an accurate diagnosis. Depending on circumstances, other testing may be required to look for possible causes of the seizure. These include blood counts for signs of infection, measurement of electrolytes (sodium, potassium, calcium) in the blood, a glucose level, and a drug screen. These tests will show if the seizure is symptomatic of another illness. If the episode is suspected to be a seizure and not due to a readily apparent cause such as low blood sugar, the physician will almost always ask for at least two tests to confirm the diagnosis and look for causes: an EEG and an MRI.

Seizure Versus Epilepsy

As we saw in the previous chapter, many conditions can look like epilepsy, especially syncope. If the physician thinks that the person has had a seizure, the next step is to determine whether the seizure was a symptom of another problem or whether the person has epilepsy and requires treatment for that. Many conditions that have nothing to do with the brain can result in seizures, so long as something has irritated the brain. Broad categories of illness include endocrine, infectious, cardiovascular, and toxic. Again, these are not epilepsy and require treatment of the underlying problem, not the seizure.

The most common endocrine problem that can result in a seizure is diabetes. Seizures can result from high blood sugar. Typically these do not happen in people whose diabetes is well controlled, but when diabetes is poorly controlled. Seizures can also happen when blood sugar is low (hypoglycemia). This does not occur with simple lack of eating, because the body is normally good at maintaining blood sugar for many days even if no food is eaten. It can happen, however, in people who take medication to lower blood sugar; they must carefully monitor their sugar to ensure that the dose is correct. Another rare condition that can result in low blood sugar is an insulinoma—a tumor that secretes insulin, with resulting abnormally lowered blood sugar.

Seizures can also occur with abnormalities of electrolytes—sodium, calcium, magnesium, or potassium. As with high or low blood sugar, this rarely happens in the absence of a disease.

Infections that can cause seizures include meningitis (infection of the coverings of the brain) and encephalitis (infection of the brain itself). These can be viral, bacterial, or fungal. Typically other signs of infection will be present, especially fever, but not always. If an infection is suspected, the physician will likely recommend a lumbar puncture (spinal tap) to look for infection in the fluid surrounding the brain.

Cardiovascular causes of seizure include arrhythmias (abnormal heart rhythms) and syncope (fainting). There will usually be other signs of heart disease, such as chest pain, pounding in the chest, or additional episodes typical of syncope without a seizure.

Toxins, including drugs, are a common cause of seizures that are not epilepsy. The most notorious drugs for causing seizures are stimulants, such as cocaine or amphetamines, and PCP (an animal tranquilizer). Seizures can also be caused by withdrawal from sedative medications, such as certain sleeping pills, painkillers, antianxiety drugs, and alcohol. Many other drugs, including some antidepressants and antipsychotic medications, have seizures as a rare side effect. A few herbal products can rarely cause seizures (including ephedra now banned in the U.S.). Finally, toxins in the environment such as organic solvents and heavy metals (mercury, lead) can rarely cause seizures.

A doctor should look into all of the above potential causes in evaluating a first seizure. In general, all of these cause a diffuse injury to the brain and therefore typically result in generalized tonic-clonic seizures. If your doctor determines that the seizure was not provoked by another cause, a workup along the lines described below is in order.

The Neurologist or Primary Care Physician

The first encounter with a physician may be in an emergency room or a private office. If in an emergency room, the person will likely be seen by an emergency medicine specialist or an internist. Depending on the hospital, a neurologist (a specialist in diseases of the nervous system) may well be called to consult. Epilepsy is a common condition; therefore all doctors have some degree of experience in its diagnosis and treatment. A consulting neurologist might be called only if the diagnosis is unclear.

Nationwide, most people with epilepsy are ultimately evaluated by a neurologist. He or she will have extensive experience with epilepsy; typically up to one-fifth of a general neurologist's patients will have epilepsy. In more rural areas, with fewer specialists, people are more likely to be evaluated and treated by a specialist in family medicine or internal medicine. As most cases of epilepsy are relatively easily diagnosed and treated, these health professionals should have no trouble treating people with epilepsy. Still, in some circumstances you may consider consulting with an epilepsy specialist; these are discussed below.

Specific Tests Commonly Ordered

EEG

Almost all people with epilepsy will need an EEG, or electroencephalogram. This test measures changes in the electricity in the brain over time. Electrodes are placed on the surface of the scalp, usually using a conductive paste. Most of the brain lies under areas covered by hair, so this is often a bit messy, but the paste washes out easily. About twenty spots are measured on the scalp, and an electrode is placed on each. There is usually also a measurement of the electrical rhythm of the heart (EKG). Sometimes, electrodes are also placed near the eyes.

During the EEG, the person is sitting or lying down. The technologist asks the person to do things—blink her eyes, repeat a phrase, answer a question. These help to distinguish electrical rhythms produced by parts of the body other than the brain (muscles and eyes especially create electrical activity that will be seen during an EEG). The doctor may want a sleep EEG, in which case the person can come to the test tired so that she is likely to take a nap, or she will be given a sleeping pill.

How does this test help in diagnosis? The brain is always producing electrical rhythms, generated by groups of neurons. The amount of electricity is small and must be recorded through scalp, bone, and the coverings and fluids around the brain, all of which decrease the signal. Some compare an EEG to trying to listen to a conversation two floors down—you need to listen carefully and will only make out words if many people shout at the same time. An EEG is generated by comparing electrical potentials at each electrode to a reference point; this can be any of the other electrodes in place. By grouping tracings from many electrodes, the technologist (and reading neurologist) can see a sort of map of the brain's electrical activity.

How can this help in diagnosis if no seizure is taking place during the test? Normal people have certain patterns of activity under various conditions—awake, asleep, concentrating on something, quietly resting with the eyes closed. Abnormalities in these rhythms may help locate a problem that could result in epilepsy. Also, most people with epilepsy will show split seconds of abnormal activity even when not having a seizure. These "spikes" (named for their shape on the EEG) probably represent a group of neurons firing abnormally, like the people downstairs shouting a single word. Spikes can be frequent, but often occur only rarely. Most are more common during sleep, which is why a sleeping test can show a problem when another done only awake does not. As spikes are random, they can show up in one test and not the next—which is the main reason EEGs are often repeated.

The EEG should be read by a neurologist specifically trained in this technique. As different laboratories and neurologists use slightly different techniques, each may want to see an actual EEG rather than the report from another laboratory.

CT Scan

A CT (computerized tomography) scan is an X-ray that has been taken from many angles and combined by a computer to create a three-dimensional image (a regular X-ray is only two dimensions: height and width). Most emergency departments have a CT scanner readily available, so this is often the first test performed to look at the brain. The images obtained are not as highly detailed as with an MRI (below), but they can easily show most abnormalities that could cause seizures: bleeding in or around the brain, tumors, and strokes. A CT scan is also much quicker than an MRI and considerably less expensive. A CT scan can assure a physician in the emergency room that no life-threatening problems are present, allowing a safe discharge home with more detailed testing (if necessary) arranged during routine office visits.

MRI

An MRI (magnetic resonance imaging) scan uses magnetic fields, rather than X-rays, to create an image of the brain. The test relies on hydrogen ions, particularly those present in water molecules in the brain. Hydrogen ions have a "spin"—each is either up or down. Under normal conditions, the spin will be randomly distributed. The strong magnet in an MRI aligns the "spin" of the hydrogen ions in the same direction. Then the magnet is turned off and each hydrogen ion drops back into a random state, but as each does so, it gives off energy. The energy is altered by the surrounding tissues and can be measured by the MRI. In this manner, the MRI creates a detailed picture of the brain. Unlike a CT scan, bone does not distort the image. The MRI may show subtle changes in certain brain areas that could be responsible for the seizures. Small vascular malformations, clusters of neurons in abnormal places, and certain types of scarring can be seen with an MRI and not with CT scans.

To generate an MRI scan, the person is placed inside a machine that has a powerful magnet. Most MRI machines are rather like a small tunnel, and people who are uncomfortable in small spaces may benefit from a small dose of sedating medication. So-called open MRIs are also available in some areas; these give the same test but are less confining. The strong magnet used in an MRI will attract any kind of metal, so all jewelry or other metal objects are removed before the test. If metal devices are in the body, such as pace-makers or clips, an MRI might be impossible.

Other Brain-Imaging Techniques (PET, SPECT, Functional MRI)

In more difficult cases of epilepsy, particularly if surgery is possi-ble, other methods of imaging might be recommended. The three described below are in common use today.

PET Scan

A PET scan (positron-emission tomography) uses radioactive compounds to create an image of the brain. The most common compound is glucose: injected, it will be taken up in each brain re-gion according to its use there as fuel. Areas using an abnormally low amount of glucose may be malfunctioning with seizures. Under certain rare circumstances, a PET scan is taken *during* a seizure. This requires a highly reliable means of provoking a seizure. In this case, the seizing area will use an abnormally high amount of glucose and be seen on the scan.

SPECT Scan

A SPECT scan (single-photon-emission computerized tomogra-phy) uses a different kind of radioactivity, one that specifically sticks to blood vessels. SPECT scans can show areas of decreased blood flow, also a possible marker for seizure foci (where the seizure starts). A SPECT injection can also be attempted during a seizure; this can show the increase in blood flow to the focus that occurs during a seizure. It is easier to obtain a SPECT than a PET during a seizure as the SPECT injection can be performed on a hospital floor, usually during video-EEG monitoring (de-scribed below), with the patient then being taken to the imaging laboratory.

Functional MRI

A special MRI called a functional MRI (fMRI) may be recommended as part of a surgical workup. This uses the same MRI machine described previously to determine the part of the brain specifically involved in certain functions, such as moving an arm or a leg. It can also be used, as is the Wada (described below), to determine which side of the brain controls language, and (to some extent) to test memory. Most do not consider this a replacement for the Wada, but this might happen in the future as techniques improve.

A functional MRI is performed exactly like a routine one, except that the person is instructed to perform specific tasks during the test such as moving, talking, or remembering. Through different magnet and recording settings, the radiologist can find areas of the brain that are working harder during the task and, therefore, are essential for performing it.

Neuropsychological Testing

Everyone has strengths and weaknesses. Some people are better at math, others at writing poetry. Some have a good memory for names, others for how to reconstruct an automobile engine. In a person with epilepsy, these strengths and weaknesses can reflect areas of the brain that are affected or unaffected by seizures.

Complete neuropsychological testing involves many tasks and typically takes a full working day to complete. Aspects of memory are tested, as are language and the ability to learn new tasks. This evaluation is nearly always performed before possible epilepsy surgery, but can be helpful in other circumstances, such as when memory seems to be worsening, or in people who are depressed or anxious.

Extended EEG Monitoring and Video-EEG Monitoring

In some cases, despite extensive testing, the diagnosis of epilepsy is unclear. In others, the diagnosis seems certain but the seizures are not responding to medications as expected. Under either of these circumstances it may be useful for the doctor to see a longer sample of EEG. A routine EEG lasts twenty to forty minutes; extended EEG monitoring can last for days or weeks.

Ambulatory EEGs can record at home for many days. The devices are highly portable; electrodes placed on the head are plugged into a small recording box that can be worn around the waist. People can walk around, work, and do most activities (except washing the hair). One of my patients gave a concert (including a violin solo) with an ambulatory EEG in place.

EEG recordings from ambulatory devices are not as detailed as the studies performed in an office, but they have two advantages. First, abnormalities may occur over the longer recording time that cannot be seen during the twenty-to-forty-minute studies routinely done in an office. As epileptic discharges are random and occur more in the deeper stages of sleep, an ambulatory EEG may clarify the problem. Secondly, particularly if seizures are happening frequently, the likelihood that an actual seizure will be recorded is higher with the EEG in place. Such a recording can often yield a definite diagnosis.

All specialized epilepsy centers have the capacity to perform video-EEG monitoring, where the person is recorded with both EEG and a video camera. Video-EEG allows the neurologist to directly observe behavior during a seizure. The simultaneous recording of EEG allows the physician to look at changes in brain waves during abnormal behavior. Recordings can be performed on outpatients, with devices that are similar to ambulatory EEG plus a somewhat bulkier camera that must be taken along wherever the person goes. However, most video-EEG evaluations are done in the hospital because this allows higher-quality recordings, testing during seizures by trained personnel, and consultation by specialists if necessary. Medications are also usually decreased or stopped to encourage seizure activity, and this is much safer when done in the hospital. Once the diagnosis is confirmed, alternative treatments can promptly be begun.

In cases where seizures happen daily or more frequently, a video-EEG evaluation can usually be completed in a day or two. But what if seizures are more rare? A person who has a seizure once a month clearly cannot spend a month in the hospital (or even at home with an extended EEG monitor). So, unlike at all other times, seizures are encouraged during inpatient video-EEG. Medication may be decreased or stopped. People may be asked to stay up late or even to drink small amounts of alcohol, which can provoke a seizure. If certain things tend to provoke seizures in a particular person (computer work, stressful games, exercise), individuals will be encouraged rather than forbidden to do these activities.

Video-EEG can sometimes show that seizures are not epilepsy at all (this occurs in about a quarter of people referred to epilepsy centers; see chapter 6). This obviously has important implications for treatment and explains why medications prescribed for epilepsy did not work. When epilepsy is present, the exact seizure type and site of onset can usually be determined. This can help in choosing another medication to try, or whether to look into other treatments such as surgery or the vagus nerve stimulator (see chapter 9). Although video-EEG monitoring typically requires days or even weeks, if epilepsy remains a problem, it can make a huge difference in solving it.

When to Consult an Epileptologist

Under most conditions epilepsy can be successfully treated by a neurologist or a primary care physician. In cases when seizures continue, even if they remain rare, people should consider seeing a specialist. A better, but more unusual, treatment may be available, or the condition might not be epilepsy at all.

Even when seizures are fully controlled, undesirable side effects from medication may occur, such as weight gain, excessive tiredness, memory problems, or (in women) excessive growth of body hair. These might disappear with equal seizure control on another medication. Women who are considering having children should consider consulting with an epileptologist, as information regarding safety of medications and recommended treatments before and during pregnancy changes rapidly.

Tests Specifically Needed for Epilepsy Surgery Evaluation

In addition to the tests described above, several others are used mainly in people who are being considered for epilepsy surgery. These include the Wada test, video-EEG monitoring with intracranial electrodes, and brain mapping.

The Wada Test

The Wada test, named for the physician who first performed it, is also called the intracarotid amobarbital test. The purpose is to determine where language function lies, which in nearly all right-handed people and most left-handed people is on the left side of the brain. It is also used to test memory on each side of the brain. The test is important for epilepsy surgery to ensure that each of these important functions will not be affected if surgery is performed. The test is performed by either a neurologist or a neuropsychologist, with the help of a radiologist. It is usually an outpatient procedure and the person returns home from the hospital the same day.

In the Wada test, a catheter (a small tube) is inserted into the femoral artery, which is the place where a pulse can be felt in the inner thigh. The tube is then threaded through the artery into the abdomen and chest and into the carotid artery in the neck, which supplies the blood to the brain. The insertion is done in the leg because this is less painful and less risky than making an incision in the neck, where bleeding carries a higher risk of damaging vital structures. Once the catheter is in place in the carotid artery (the radiologist can tell by a type of X-ray called fluoroscopy), dye is injected through the catheter and X-rays are taken in rapid succession, showing the flow of the dye through the arteries and veins of one-half of the brain (an angiogram). An EEG is usually done throughout the test to measure brain-wave activity.

After this, another injection is performed using a short-acting anesthetic, usually sodium amytal. This will effectively put half of the brain to sleep for about five minutes, and the other half can then be tested. While the anesthetic is working, the person will be paralyzed on one side and may be unable to speak or may speak abnormally. A number of pictures or objects are then shown to the person, who may not be able to say what they are. Regardless, the person will be asked to remember what they are. Once the anesthetic wears off, the neurologist or neuropsychologist will ask the person what she remembers.

If a part of the temporal lobe, the hippocampus, needs to be surgically removed to treat epilepsy, this test assures that memory on the side that will not be operated on works well. Also, if memory on the surgical side is poor, it suggests that this area is generating seizures and is not being heavily relied on for memory. Hence, there is less of a chance of a problem if it is removed. If on the other hand memory on the surgical side is also good, the chance is greater that the person will have some memory changes after the operation, particularly if the language side (the dominant hemisphere, usually the left) is the side that must be operated on.

As with all tests, the Wada must be interpreted in light of all the other information from other tests. In itself, this test cannot predict exactly where the seizures are coming from or whether there will be problems after surgery, but it is an important part of the information needed for a recommendation on surgery.

Video-EEG Monitoring with Intracranial Electrodes

If surgery for epilepsy is being considered, the neurologist and the surgeon must be convinced that they know exactly where the seizures are coming from. They must also know that the area can be safely removed without causing a significant problem. All of the other tests listed above can be used for this purpose, and often they will be enough to comfortably show what the required surgery should be. When the area is not clear, however, the person may need a further test to show precisely the site of seizure onset: a video-EEG recording with intracranial monitoring, that is, with electrodes placed directly on the surface of the brain.

After performing all of the other tests needed for surgery, the neurologist will almost always have a fairly good idea of where the seizures start. This would be the area targeted with intracranial monitoring. This procedure begins with an operation, sometimes with holes drilled in the skull and electrodes slipped through them to record a relatively small area of brain. Other people need a larger area of brain recorded; in this case, the surgeon will open the skull and the outside coverings of the brain and place the electrodes directly on the brain's surface. This allows a much clearer recording than with electrodes on the scalp. After the surgery, the person awakens from anesthesia and is sent for video-EEG monitoring. This recording takes place otherwise much like scalp monitoring: medications may be lowered to bring on seizures, and enough seizures are recorded to show exactly where they come from. The electrodes are then removed by the surgeon, and sometimes the actual epilepsy surgery (removal of the abnormal brain area) is performed at the same time.

Sometimes, all of the testing done before intracranial electrode placement does not even show on which side of the brain the seizures begin. In these cases, if the person is otherwise a good candidate for surgery, electrode strips can be placed on both sides of the brain through holes drilled in the skull. Recording of seizures with these in place will usually help the surgeon and epileptologist narrow down the region of onset. Most of these persons will then need another operation in which electrodes are placed more precisely in the area of seizure onset, as described above, before epilepsy surgery.

Placement of intracranial electrodes carries some risk, although small. The operation can cause bleeding into the brain or an infection of the brain, particularly if the subsequent recording is prolonged. Because of these risks, intracranial electrode recordings are usually performed only after all other tests have been done and have failed to give enough information to direct epilepsy surgery.

Brain Mapping

In "brain mapping," specific spots on the brain's surface are tested to determine their function, then all the information is put together into a "map" of the brain. This is often done during an intracranial recording, using the electrodes that have already been placed on the brain's surface. It can also be done during surgery, when the brain is exposed.

During brain mapping, the physician will deliver a small electrical shock to an area of the brain (through the external wires during an intracranial recording, or directly on the surface during surgery) and will then observe the result. If the part controlling the left arm is stimulated, for example, the arm will move (even if the person is asleep). If the person is awake, he can describe what happens: a flash of light, a tingling in the right toes, even his typical aura. The location of language can be tested in this way also, particularly if the information from the Wada was not sufficient. If the testing is performed during the operation itself, the person may need to be awake. This can be done with little discomfort; the brain does not perceive injury to itself, and the rest of the operation (opening the scalp and skull) can be done using local anesthesia along with mild sedatives. The information from brain mapping is particularly important if the operation is being performed in areas close to those controlling important functions, such as movement or language, and can direct the surgeon to avoid them.

All of the techniques described in this chapter are meant to aid the physician in finding the best treatment possible for each person with epilepsy. The chapters in part 2 describe the treatments themselves, all meant to give the highest chance of complete freedom from seizures.

PART TWO

Treating Epilepsy

8

The Mainstay of Treatment: Many Medicines for Epilepsy

As with many treatments in medicine, the first drug for epilepsy was tried serendipitously. A family of drugs, the bromides, had been used primarily as sedatives and to control excessive sexual urges. A popular belief in the mid-nineteenth century was that epilepsy could be caused by masturbation; therefore, it was reasoned, bromide would decrease the sexual urge and improve the epilepsy. The treatment worked, but for totally unrelated reasons. In other cases, bromides were administered to people to treat psychosis (hallucinations and paranoid thinking, most closely associated with schizophrenia) and the seizures incidentally improved in people who also had them. Over the second half of the nineteenth century, bromides became the fifth most popular drug prescribed by physicians in the United States. They fell out of use when another drug, phenobarbital, became available and had fewer adverse effects.

For people who are diagnosed with epilepsy, a doctor will almost certainly recommend treatment with an anticonvulsant drug (also called an antiepileptic drug, or AED). Although it may seem annoying to take a drug several times a day to prevent an event that may happen only once a week or once a year, any small risk associated with taking a drug is almost certainly less than the risk of repeated seizures (see chapter 13). There are a few exceptions. If seizures are mild and the person remains fully alert during them, it might be reasonable to monitor the person carefully without prescribing medication. If seizures occur because of some other illness (such as meningitis or encephalitis), the person may not need treatment once the illness resolves (and by definition this is not epilepsy, but "symptomatic seizures"). Sometimes mild seizures in children are not treated if they are not old enough to drive and if they can be closely supervised to avoid injury. Some benign types of epilepsy in children will eventually stop, and these are sometimes not treated. But by and large, the vast majority of people need to take medication regularly.

A large number of drugs are available, and at first glance distinguishing between them can be confusing. Communication with your doctor is essential in choosing the correct medication for you. Many medications take time to reveal whether you experience adverse effects, and your partnering with your doctor will help to ensure that you remain comfortable with taking your medication daily.

The two broad categories of anticonvulsant drugs are those that treat only partial seizures, and those that are broad-spectrum and treat nearly all types of seizures. Within these categories, clear differences in effectiveness between drugs have never been proven: one is just as good as the next (at least as far as studies show). For the most common seizure type, partial seizures, any drug has about a 70 percent chance of being effective. For reasons that are unclear, in one individual a certain drug will be completely effective while another will not. Doctors may have (a sometimes strong) opinion that a particular drug works the best overall, but objective data are simply not there: when various drugs are compared in carefully designed trials, none is definitely more effective. There are significant differences, however, in other properties of anticonvulsant drugs, such as side effects, ease of use, and cost. These differences can be important, as the wrong match between person and medication can make the treatment burdensome and sometimes worse than the disease itself. These properties are therefore important in choosing the first drug tried, and the "best" drug will be different in different people. In the 30 percent or so of people in whom the first drug selected does not work, it makes sense to try another—but again, there is no concrete evidence that one drug is most likely to be effective. Please note that throughout this chapter generic drug names are typically used. If you need to check this against your brand, refer to Table 8.1 at the end of this chapter.

How Do Drugs for Epilepsy Work?

Epilepsy drugs actually have many different actions in the brain. All, however, have the potential to affect the transmission of abnormal signals between neurons—the cause of seizures. Simply shutting down all transmission among the malfunctioning neurons would clearly stop seizures, but this would also shut down all normal brain processes associated with those neurons as well. The trick is to calm the abnormal activity without affecting normal thinking and alertness.

Many drugs that are effective for partial seizures affect a specific chemical channel of the neuron: the sodium channel. This is a "gate" through which sodium ions can rush into the cell and is needed for the neuron to fire. Drugs such as phenytoin, carbamazepine, and lamotrigine affect this channel to prevent neurons from firing rapidly (as during a seizure). Other drugs, such as valproic acid, ethosuximide, and levetiracetam, act on calcium channels (also involved in neuronal firing) in a similar way. Phenobarbital and and benzodiazepines act through a specific type of neuron that uses gamma-aminobutyric acid (GABA) as its transmitter. GABA neurons are usually part of an inhibitory system, and drugs affecting GABA probably work by improving the brain's self-calming ability to quell seizures. Still other drugs, such as topiramate and felbamate, decrease excitatory pathways (mainly through systems that use glutamate as a transmitter). All drugs are better at preventing the spread of seizures than at preventing their initiation, so that small seizures (fortunately the least disruptive) are sometimes the most difficult to completely eliminate.

Although there are theories, it is not known precisely how these drugs work. Some actually seem to have several actions that could potentially calm seizures; others do not have any of these known actions yet are still effective anticonvulsants. As the mechanisms of the drugs and the precise abnormality in individual people are not well understood, finding the right match between medication and person is sometimes a process of trial and error.

What Are the Side Effects?

Since all drugs for epilepsy work on the brain, all have the potential to affect alertness and thinking, but this usually occurs only at high doses. Some drugs are more likely to do so than others. Phenobarbital, for instance, is one of the most sedating of the epilepsy drugs. It is in the same drug class (barbiturates) as many sleeping pills. Some evidence suggests that carbamazepine and phenytoin affect learning and memory more than other drugs, such as gabapentin and lamotrigine, although the differences are small and (in most people) insignificant.

Another relatively common side effect is nausea and stomach

pain (more common with valproic acid, tiagabine, zonisamide); this usually improves if the medication is taken with food. Cognitive (thinking) problems (including difficulty speaking, memory problems, and concentration problems) are sometimes seen with topiramate. Weight gain can occur with valproic acid and gabapentin; weight loss with felbamate, topiramate, and zonisamide. Valproic acid can cause a syndrome of cystic ovaries, weight gain, and excessive body hair growth that may be a result of changes in female hormones. Zonisamide and topiramate involve a small risk of kidney stones.

Many other side effects are possible. In the package insert of each drug, anywhere from dozens to hundreds of possible adverse reactions are listed, but typically these are quite rare and may occur in less than 1 percent of people.

Are There Serious Risks with Anticonvulsants?

The side effects listed above are mild and will stop if the drug is stopped. Some drugs are associated with rare, more serious side effects. Felbamate, for instance, can rarely cause aplastic anemia, a potentially fatal condition in which the body stops making blood cells. Although unusual (probably about 1 in 5,000 chance at most), this is a more common occurrence than with other drugs and is the main reason felbamate is rarely used. The same reaction can occur with phenytoin or carbamazepine, but probably at most about 1 in 100,000 people are affected. Any drug can be associated with an allergic reaction, which usually begins with a rash but can become severe in rare cases. Rash is unusual with valproic acid, gabapentin, or levetiracetam. A few drugs also have a rare complication of liver failure. This is most problematic with valproic acid used in young children. Finally, certain drugs may slightly accelerate osteoporosis.

This is most likely to occur with phenobarbital, phenytoin, valproate, or carbamazepine. It is more problematic in women, as they are already at higher risk of osteoporosis than are men.

This may sound as if a lot of terrible things can happen as a result of taking anticonvulsants, but in general these are safe medications. Keep in mind that the risks of the drugs are extremely small and (in most cases) much less than the risks from continued seizures.

How Long Must the Drug Be Taken?

Usually, anticonvulsants must be taken for at least several years. It is sometimes reasonable to try stopping them after being seizure-free for two or more years. This is another individual decision, though. With a known abnormality in the brain (such as damage from a stroke) or a persistent abnormality on EEG comes a higher risk of the seizures returning. Also, certain types of epilepsy are known to persist, even when the seizures are easily controlled. Anticonvulsants should never be stopped without consulting a doctor, no matter how confident the person feels that the condition is improved.

What About Generic Drugs?

For many conditions, generic drugs are a reasonable, less costly option for treatment. For epilepsy, however, use of generic drugs is usually not advisable. The main problem lies in differences in "bioavailability," or the amount of drug that is actually absorbed into the body. Bioavailability can be affected by the amount of drug in the tablets or capsules and their method of preparation, among other things.

The Food and Drug Administration allows generic drugs to vary by 20 percent (up or down) in bioavailability when compared to a brand-name drug. That means that when taking a 100 mg tablet, anywhere between 80 and 120 mg may actually appear in the system. This is a considerably wider range than the amount a brand-name medication is allowed to vary (only plus or minus 5 percent). A difference of 20 percent in the formulation of a pain medication, for instance, may not significantly change its effectiveness. But treatment of epilepsy, unlike most conditions, requires the person receive a fairly specific dose of a medication. Too much and she will likely become tired or dizzy. Too little and she will be at risk for seizures. The biggest problem comes when switching between different manufacturers of the same generic drug, which pharmacists commonly do when a prescription is filled, and the change is frequently not apparent to the consumer. Changing from one generic preparation to another could potentially result in a 40 percent swing in the amount of drug in the body. That's why, in general, most neurologists recommend a brand-name drug for treating epilepsy.

Brief Descriptions of the Most Commonly Used Older Drugs

Phenobarbital

Phenobarbital is in the class of drugs called barbiturates, which are also used as sleeping pills, so it is probably not surprising that drowsiness is one of its most common and troubling adverse effects. Phenobarbital was initially tried for similar reasons as bromide. Over time, most people develop tolerance to phenobarbital, meaning that they are no longer as tired. Their bodies have become accustomed to the drug, so much so that if it is stopped, they will experience an opposite effect—they'll feel nervous, shaky, and unable to sleep. This reaction is identical to what is termed addiction with drugs of abuse, but since phenobarbital is therapeutic, this change is sometimes called dependence.

Phenobarbital is prescribed less often today but is still a useful drug for partial seizures and generalized tonic-clonic seizures. One advantage it has over some other drugs is that it can be taken once a day, and it is by far the least expensive medication for epilepsy. For this reason it is more commonly used in developing countries. Besides sedation, other common side effects are dizziness, clumsiness, and blurred vision. Rash can occur, and this can rarely lead to a serious reaction. Other medications must be taken cautiously with phenobarbital because of the risk of interactions (particularly other sedatives). Phenobarbital can increase the risk of osteoporosis.

Mysoline (primidone) is a drug similar to phenobarbital with similar effectiveness and side effects. It is actually metabolized in the body into phenobarbital, but primidone itself is also an effective anticonvulsant.

Phenytoin (Dilantin)

Phenytoin was the next major drug introduced after phenobarbital, in the 1930s. Phenytoin was designed as the first epilepsy drug that was not a sedative. It was developed through the cooperation of scientists at Harvard University and a pharmaceutical company, Parke-Davis (now part of Pfizer). Phenytoin is much less sedating than phenobarbital or bromides, although it has other side effects. Still the most prescribed anticonvulsant in the United States, phenytoin has slowly been being replaced with drugs with fewer side effects.

Phenytoin is still prescribed because it has been in use for over half a century and physicians are comfortable with it, and because it can be started quickly in an emergency room (by mouth or intravenously). However, it is probably not the best drug now available. Its most common side effect is tiredness. Dizziness and unsteadiness can occur at higher concentrations. Rash can occur and may rarely lead to a serious reaction. Phenytoin has complicated interactions with other drugs, such that special care must be taken when it is combined. The difference between an effective dose and a toxic one is small and can vary even within the same person, making toxicity more common than with other drugs. Oral contraceptives are less effective when a person is taking phenytoin. More problematic (and responsible for its decline in use) are long-term side effects. These include gum thickening (sometimes requiring surgery), "hirsutism" or masculinization (increased hair growth and coarsening of the face, particularly troublesome in women), increased osteoporosis, and neuropathy (a nerve problem resulting in decreased feeling, usually in the legs). Nevertheless, it is very effective for partial seizures and most people will have no problems taking it. In most people, it can be taken in a single dose daily.

Carbamazepine (Tegretol, Tegretol-XR, Carbatrol)

Carbamazepine is another drug used exclusively for partial seizures. Worldwide, it is actually the most prescribed drug for seizures. Carbamazepine was first used to treat another condition, called trigeminal neuralgia or tic douloureux, a painful irritation of the nerves in the face. It is still used for this, but prescriptions for epilepsy far exceed this use. In general, carbamazepine is less sedating than phenytoin, and it does not cause hair growth or gum thickening. Some reports suggest that it is less likely to cause problems with memory and learning than phenytoin (although this is quite rare with either drug). Rash is not uncommon (about 10 percent of people will experience this), and it can rarely lead to a serious reaction. Carbamazepine can interact with other drugs, including oral contraceptives; therefore, it is always important to check with your doctor or pharmacist before taking another drug (even over-the-counter or herbal preparations).

Carbamazepine seems to have effects on neurons similar to those of phenytoin. Since the two drugs are far from identical, this shows that our knowledge of how these drugs actually work is probably incomplete. People can have side effects on one and not the other, and others may have seizures fully controlled on one drug but not the other. Carbamazepine is usually taken in three or four doses per day, but the extended-release forms (Tegretol-XR and Carbatrol) can be taken in two doses.

Ethosuximide (Zarontin)

Ethosuximide is unusual in that it is useful only for absence seizures (brief staring spells occurring primarily in children). It will have no effect on partial, myoclonic, or other seizure types. It has few side effects at the usual concentrations, although tiredness, dizziness, and stomach upset have been seen. Because absence seizures are much more common in children than adults, ethosuximide is more commonly taken by children.

Valproate (Depakote, Depakote ER, Depakene)

The anticonvulsant properties of valproate (including valproic acid and divalproex sodium) were discovered accidentally during the screening of many other drugs for use in epilepsy. These investigators were using valproic acid not as an anticonvulsant, but to dissolve the other drugs. When all of the drugs worked, they determined that valproic acid, the solvent, had the beneficial effect. Slightly different preparations of the same drug, including divalproex sodium, were subsequently developed and together are referred to here as valproate. Valproate was the first true broad-spectrum drug for epilepsy, and it treats virtually all types of seizures. Development of valproate in the 1970s resulted in the first effective treatment for many people with seizure types other than partial.

Valproate has some actions similar to carbamazepine and phenytoin—it slows the firing of neurons. But it seems to have other actions as well, particularly through calcium channels used to control neuronal firing, and through an inhibitory neurotransmitter, GABA. Like phenytoin and carbamazepine, it can cause tiredness, but this is unusual. More common are stomach pain and nausea; the Depakote preparation is actually a coated tablet to help prevent this and is often better tolerated. Other troublesome side effects are weight gain and tremor. Rash is unusual with valproate.

Valproate has a different effect on other drugs from phenytoin and carbamazepine. It actually makes the liver less efficient at breaking down some other drugs. This results in insignificant changes in nearly all drugs, but another epilepsy drug, lamotrigine, is sensitive to this effect, and people who take both drugs together need a much lower dose of lamotrigine than they would otherwise. The effect may be important, however, in the regulation of the body's natural hormones. In some women, changes in hormones due to valproate may contribute to a syndrome including cystic ovaries and masculinization. Overall, however, valproate is well tolerated in most people. It is usually taken in two daily doses, but the extended release form (Depakote ER) can be taken in one dose.

A Success Story

Here is an example of how the "older" group of epilepsy drugs just described can successfully be used. Bob, an investment banker, was thirty-five years old when he had his first seizure. The only unusual thing he recalls about the day of the seizure was that he stood up suddenly in a moving van, striking his head on the roof. He didn't pass out, and this almost certainly had nothing to do with the seizure he had that night. He remembers nothing at all of the seizure: he went to sleep peacefully and woke up hours later in the emergency room. His wife, however, witnessed the seizure. She awoke to find his entire body stiff and his face turning blue. Within thirty seconds, he began to have violent jerking of his arms and legs. He bit his tongue, and she was terrified to see blood oozing out of his mouth. After what seemed like an eternity (but was actually only about ninety seconds) his body relaxed, with loud snorting breaths, but he was completely unconscious for another five minutes. The ambulance arrived as he was just beginning to awaken and to mumble incoherently. By the time he arrived at the emergency room, he was awake and appeared normal. Blood tests were normal, and a CT scan of his brain and an EEG were also normal. Taking a careful history, the ER physician found that about a month earlier Bob had awakened in the morning feeling sore, with his tongue bitten and the sheets in disarray. His wife had been away on an overnight trip, so it wasn't clear what had happened, but in retrospect it seemed clear that this was almost certainly another seizure.

The treating physician wanted to make sure Bob would not have another seizure when he was released that morning, even though the chances were small (probably less than 10 percent over the next couple of days). Carbamazepine was available in the emergency room, but the dosage must gradually be increased over about one week or people become dizzy and nauseous. The physician felt that valproate had more side effects than phenytoin, although both could be started intravenously. He decided to prescribe phenytoin, initially given intravenously. This made Bob unsteady on his feet, but he had a fully effective amount in his system before leaving the emergency room. (If his patient had been a woman, the physician might have chosen another drug because of the common cosmetic side effects of phenytoin.)

After three days of taking phenytoin by mouth, Bob felt completely normal. He did not develop gum problems and did not notice any other problems. As he was otherwise healthy, he rarely took other medications, so drug interactions were not problematic. After two years with no further seizures and a normal EEG, the phenytoin was stopped and Bob never had another seizure.

Drug Development and the New Generation of Epilepsy Drugs

The drugs listed above (along with a few others that are now rarely used) were the main treatments for epilepsy in the 1970s and 1980s. In 1990, no new medications for epilepsy had been introduced in fifteen years. Sometimes none of the available drugs were satisfactory in a given person. The reasons were variable. Many people found the drugs (particularly phenobarbital and primidone) sedating. Even though they might have no seizures, people slept ten or twelve hours a night and still felt tired during the day. Phenytoin is less sedating, but frequently causes cosmetic effects: almost everyone eventually develops thickening of the gums, although it may take years; some people need gum surgery twice a year. About 30 to 40 percent of people still had seizures while taking medication.

In the late seventies and early eighties, the National Institutes of Health (NIH) recognized the need for new seizure drugs and began a program of aggressive screening of possible agents. Researchers already knew that a number of animal models of epilepsy could predict whether drugs would be useful in people with seizures. Most involved rats or mice that either were born with seizures or were given seizures by injection of a toxic drug (pentylenetetrazol) or by repeated electroshock. The NIH made it possible for anyone—drug companies or university-based researchers—to submit drugs that would then be tried in these models.

The screening of hundreds of drugs like this, hoping that some will be effective, might seem rather crude. But because it is not really known how drugs for seizures work, this is the best method available so far. Ideally, the precise problem in a person would be identified and a drug designed for that, but at present that is not possible.

The NIH program was successful. Drug companies sent hundreds of different agents to be screened. The successful ones were then investigated for toxicity. If they passed this stage, they were tried in human volunteers and, finally, in people with epilepsy.

The process of taking a drug from the chemistry laboratory to the pharmacy shelf is much more involved and complicated than even most physicians realize. Suppose you believe that a drug chemically similar to carbamazepine might be more effective against epilepsy. You might have specific ideas about how to improve on it: change one part of the molecule, for instance. You turn this idea over to chemists, who synthesize several compounds with that idea in mind. These compounds are then screened in animal models of epilepsy.

If your compounds work in a variety of animal models, then they must be tested for toxicity in animals, by giving them large amounts of the drug. One may cause liver damage in the animals; another may make them so sick they will not eat. But one or two may be well tolerated and not apparently toxic. The first test in humans is usually with volunteers, who agree to take a single dose of the medicine and have a number of subsequent tests done. Drugs that have no effect in animals rarely cause problems in humans at this stage, but clearly the investigation must be done carefully. If one subject develops an abnormality—a change in a liver test or heart rhythm, for example—the drug may not be investigated further, or a number of other tests must be undertaken to determine whether the change was really related to the drug. If the drug passes muster at this level, it then proceeds to carefully designed trials in people with epilepsy. In medicine, the clearest results are always in "double blind" trials—where some people will receive the drug and others get a placebo or dummy pill, and neither the researchers running the trial nor the people taking the drug know who is getting what. Since most people with epilepsy need a drug, placebo trials are not usually performed in people with epilepsy unless they continue to have seizures while on other medications. In that case, they may be willing to try adding either drug or placebo (for a short time) to test the drug.

Only a drug that passes each stage of testing will reach the pharmacies. Most estimates place this at one drug in a thousand or so, and the cost is in the millions of dollars.

Largely because companies had the ability to screen epilepsy drugs through the NIH, many were investigated during the 1980s. Eight were approved by the FDA between 1993 and 2000. None is perfect, but in any individual, each has the potential to work better than older medications. Largely because of the way drugs are tested for epilepsy, the initial FDA approval for virtually all of these drugs was only for partial seizures, and only for use in combination with other epilepsy drugs. Since their approval, however, abundant evidence supports their use alone for epilepsy, and many (as described below) have been found to be effective for other seizure types.

Another, More Complicated Case: Part One

Many illnesses may seem like a bad dream, but Manisha's really began as one. She was fourteen, asleep in her bed. She suddenly sat up with her eyes wide open, panting as if she had just sprinted a mile. Her sister asked her, "What is it?" Manisha just stared. After two minutes, her eyelids slowly floated down. She lay back down and was once again asleep.

These strange "nightmares" happened weekly. While her family (and Manisha, who had no recollection of these) thought them no more than nightmares, a family member who was a neurologist suspected something else: "You need to see a doctor." A year went by, and the nightmares continued, but Manisha's mother was sure she would grow out of them. Then in school one day while she was walking down the stairs, Manisha had a sudden, intense feeling unlike any she had had before. It was as if everything were strangely familiar, as if she knew exactly what would happen next. Then she passed out and fell to the floor, eyes still open. She told her family what had happened and was admonished not to try so hard to get attention. These episodes happened several more times over the next months. Manisha didn't tell anyone about the feelings; she thought they were too weird and she didn't fall again. She began to think maybe she was going crazy. But the rest of the time, she felt quite normal. Still, she didn't want her parents taking her to a psychiatrist.

One day while she was writing notes in math class, she felt the strange feeling come over her again. She became distracted, looking around to see if anyone else was behaving strangely. She'd never felt it this intensely: it was as if her most emotional memory was wrapped up in what was happening at that moment. She wasn't sure if she lost touch with reality for a few moments. Her teacher saw her stare at the wall, then slowly topple out of the chair onto the floor. The school nurse was called, who found her on the floor, staring. Again, her lids dropped and she awakened, asking what had happened. Still, she wasn't taken to a doctor. Finally, it happened while the family was eating dinner. On came the same feeling, and by this time Manisha had the forethought to put down her fork. It will pass, she thought. Her mother asked if she was okay. The next thing she remembered was waking up on the floor, with emergency medical services attending her.

Only much later did her parents tell her what had happened. As the feeling had grown stronger, she had not answered her mother but stared straight ahead. Over the next minute, her expression changed from empty staring to a frightening contortion. The left side of her face began to twist, then her arm stiffened. Finally, as she fell out of the chair, both arms and legs were shaking violently as her parents frantically called 911. Within a minute, everything stopped and she lay motionless, breathing hard. To her terrified family that minute seemed like hours.

All of the episodes that Manisha was having were seizures. The intense familiar feeling, or déjà vu, is common with seizures beginning in the temporal lobe, an area involved with emotion. Some people also experience intense fear, or pleasure, as the seizure begins. Manisha's first seizures were relatively small, and because they stayed confined to one area of the brain, she was mostly able to keep control of herself and to hide the episodes from others. As is common with partial seizures, however, she eventually had a secondarily generalized, or grand mal, seizure, which spread to involve her entire brain.

Manisha underwent the usual tests at the hospital: blood tests to make sure the seizure wasn't caused by a problem with glucose, sodium, or calcium levels; a thorough examination to look for other signs of neurological damage; a CT scan of her brain to look for a tumor or evidence of a stroke. Unfortunately, her doctors did not recognize that her previous, more subtle episodes had also been seizures. Finally the family member who was a neurologist actually saw one of the episodes. She immediately recognized that this was a seizure and sent her to another neurologist, where the diagnosis was confirmed.

The doctor first prescribed phenytoin. Manisha took three hundred milligrams every day. At first the medication made her sleepy, but after a week or so she began to feel better. Although she didn't have any more grand mal seizures, the smaller ones continued. Like most medicines for seizures, phenytoin works by preventing the seizure's spread, so continuation of smaller seizures is not uncommon. Now that her parents knew what was happening, it became clear that Manisha wasn't fully aware even during the small seizures she thought she could hide. She couldn't be sure, but it seemed as if a small seizure was happening at least every week. Over the next six years she took a number of other medicines to try to control the small seizures: carbamazepine (Tegretol) worked no better, although she felt a little less tired; lamotrigine (Lamictal) gave her a severe rash, so she had to stop taking it; and valproic acid (Depakote) didn't seem to work at all.

Because she had taken so many drugs without improvement, surgery was considered. She was admitted to a hospital's Epilepsy Monitoring Unit, so that her seizures could be recorded on videotape and with simultaneous EEG. The seizures looked temporal, but the recording was not clear enough to confidently recommend surgery. Typically, other tests may confirm the area of seizure onset. MRI is the most important of these, but Manisha's was completely normal, meaning she would probably need electrodes placed on the surface for her brain to look for where the seizures precisely started. She didn't feel ready for this. She wanted to try another medicine, but not many options were left. I saw her toward the end of her hospitalization, when she and her doctor had decided to try another medication and wanted advice about which one to begin.

The Newer Epilepsy Drugs

Felbamate (Felbatol)
Felbamate was the first new drug for epilepsy in fifteen years, approved in 1993. It was released with much fanfare as a highly effective drug with no serious side effects. While older drugs tended to

be sedative, felbamate was somewhat of a stimulant. While some older drugs caused weight gain, felbamate resulted in weight loss. And while the most popular of the older drugs (carbamazepine and phenytoin) worked only for partial seizures, felbamate worked for all types of seizures. Older drugs rarely caused liver failure, low blood counts, or low sodium, requiring periodic blood tests; felbamate did none of this in the early trials and so was marketed as needing no blood tests.

A huge need existed for a new drug at this point. Hundreds of thousands of people were not fully controlled. Many others had no seizures, but were crippled by side effects. In less than a year, over 100,000 people were placed on felbamate.

In early 1994, a problem became evident. A few people taking felbamate developed aplastic anemia: the inability to produce new blood cells. It is frequently fatal. The disease occurs rarely in the general population, and a bit more frequently in people taking some of the older drugs for epilepsy. By the summer of 1994, the disease was occurring more frequently in people taking felbamate than would be expected by chance, and every physician in the country received a letter about the potential complication. Felbamate acquired a "black box" warning, so called because it appears within a box in the *Physicians' Desk Reference* (which contains the official prescription-drug information approved by the FDA). Because of this new warning, most people, even those who had done well with felbamate, were taken off the drug while the true risk was determined. Shortly after the warning about aplastic anemia, several people developed liver failure while taking the drug. Again, this occurs rarely with other drugs for epilepsy, but the true risk with felbamate was not known because the drug was so new.

Since that time, more has been learned about the possible complications of felbamate. The complication of aplastic anemia seems to be more common in people with disorders of the immune system. It may be less common in children and has never seen after the person has taken the drug for over a year. Many of felbamate's advantages are still true: it is one of the least sedating medicines and is sometimes effective when others are not. It tends to cause weight loss, an advantage for many people (although a disadvantage for some). It has other, less serious side effects in some people: insomnia, nervousness, nausea—but most people feel quite good taking it. Because the risk of serious complications remains higher than with other epilepsy drugs (estimated at between 1 in 5,000 and 1 in 25,000 people, compared to about 1 in 100,000 for most drugs), felbamate is usually tried only after a number of other drugs have failed.

Gabapentin (Neurontin)

Shortly after felbamate was introduced, another drug, gabapentin, hit the market. Gabapentin's molecular structure was similar to that of GABA (gamma-aminobutyric acid), one of the major inhibitory neurotransmitters. It made sense that an overactive brain, having seizures, might do better if it was given more of this calming transmitter. It turns out that it works differently, though a specific calcium channels, thereby decreasing new transmitter release. However, it is still effective for epilepsy.

Gabapentin is very different from felbamate. Rather than working in all seizure types, gabapentin works only for partial seizures. Rather than having many interactions with other drugs, gabapentin is one of the few that has absolutely none. But perhaps the biggest difference is that gabapentin is one of the safest drugs for epilepsy. Although it has been taken by literally millions of people, there have been no serious complications.

Gabapentin also has few less serious side effects, and most people have no problem taking it. It rarely causes sedation. A few people gain weight. Most people can start gabapentin quickly without problems, even taking a full dose the first day if necessary, although typically dosage is increased over several days. The main problem with gabapentin is that, because it does not last long in the body, it must usually be taken three or four times a day—inconvenient for many people. It also has "dose-dependent absorption"—the proportion absorbed into the body becomes smaller as larger amounts are taken at the same time. Taking smaller doses several times per day therefore also increases the overall amount of drug getting into the body.

Gabapentin is also being used to treat conditions other than epilepsy. It is effective in treating certain types of chronic pain (ultimately receiving an FDA approval for this as well), and it is sometimes used for headache. It is used in certain psychiatric conditions, particularly anxiety disorders. It is such a safe drug that it is often the first drug tried, both for epilepsy and for these other conditions.

Lamotrigine (Lamictal)

Lamotrigine was introduced in 1995. Like felbamate, it seems to work well for many types of seizures although rarely it worsens myoclonic seizures. Like gabapentin, it does not affect the metabolism of other drugs, although a few drugs will affect the metabolism of lamotrigine. Lamotrigine has no long-term toxic effects, and most people tolerate it well. It is usually not sedating; rarely, in fact, people taking it develop insomnia. Often, these are people who formerly slept nine or ten hours a night on another more sedating drug who now felt well sleeping for only seven hours nightly. This was not insomnia, but returning to normal!

Like gabapentin, lamotrigine has no known effects on the metabolism of other drugs and does not affect hormones, including oral contraceptives. It has been FDA-approved for depression seen in bipolar disorder, and some studies also suggest that it can help with the depression commonly seen in people with epilepsy.

One problem with lamotrigine is also common to many other drugs: rash. Young children are probably more susceptible to this, as are people who are taking valproate with lamotrigine. If the person continues to take the drug, the rash can turn into a serious allergic reaction, requiring hospitalization and in a few cases resulting in death. When the drug is started carefully, usually over six weeks, rash is extremely uncommon if not unheard of. No other serious problems have been seen in people taking lamotrigine.

Topiramate (Topamax)

Topiramate is another broad-spectrum drug used for all types of seizures. If started too quickly, it causes some people to feel tired and dull, even to the point where they have difficulty functioning at work. For this reason topiramate (like lamotrigine) is started slowly, usually over about six weeks. Even then, mental dullness can occur, particularly at higher doses or when taken along with other epilepsy medications. Other than this, side effects are similar to other drugs, including tiredness, dizziness, and unsteadiness. Some people lose weight while taking topiramate, an advantage to most, but unhealthy for a few people who may already be underweight. Topiramate slightly increases the risk of kidney stones, and people should be sure to drink plenty of fluids when taking it. Glaucoma can rarely occur.

Like lamotrigine, topiramate has few effects on other drugs. The main exception is oral contraceptives: topiramate can decrease their effectiveness, resulting in a greater risk of unplanned pregnancies. Topiramate is being used to treat migraine as well as epilepsy.

Tiagabine (Gabitril)

Tiagabine was released in 1997. Probably effective only for partial seizures, it works by increasing GABA, although by a different mechanism than gabapentin. It can cause sedation and stomach problems, and it, too, is usually started over six weeks. It lasts as little as two hours and so must be taken frequently during the day.

Tiagabine is not used as frequently as other newer anticonvulsants. It was effective in clinical trials, but the combination of more side effects (chiefly drowsiness and stomach discomfort) than other drugs, and the need to start slowly and take it frequently, has made it used less often. Tiagabine does not seem to affect other drugs.

Levetiracetam (Keppra)

Levetiracetam is a broad-spectrum anticonvulsant and works for partial seizures and for all types of generalized seizures. It is a "clean" drug: like gabapentin, it has no known effects on other drugs. Most people can tolerate a full starting dose on the first day of treatment, as sedation is rare. In people whose seizures are poorly controlled, this can be an advantage over drugs like lamotrigine, topiramate, and tiagabine, which take weeks to get to an effective dose. As with other drugs, some people taking levetiracetam can feel sedated or tired, and occasionally people develop anxiety or depression. Rarely, people can become paranoid or develop other psychiatric problems as a side effect of taking levetiracetam. No long-term toxic effects have been seen, making this one of the safer drugs. Levetiracetam is a relatively new drug, but has now been taken by enough people that new serious problems, even if rare, are unlikely to be found.

Oxcarbazepine (Trileptal)

Oxcarbazepine is chemically similar to carbamazepine, but because it is metabolized differently, some people may have fewer side effects on this drug. Some people can be controlled taking it twice a day, but many need to split up the dose, taking it three times a day. It may be particularly useful in people whose seizures improved on carbamazepine but who had side effects (usually tiredness or double vision) on that drug. Toxic effects overall are similar to those of carbamazepine: oxcarbazepine can cause lowering of blood counts or sodium, although these are typically not problems. It also has some effects on other drugs, including oral contraceptives, although these are less pronounced than with carbamazepine.

Zonisamide (Zonegran)

Zonisamide was largely developed in Japan. While relatively new to the United States, it has been used in Japan for ten additional years. It has several of the potential advantages of the other new drugs. Zonisamide works in all types of seizures and is widely used for myoclonic and primary generalized seizures as well as partial seizures. It can be started at an effective dose on the first day of treatment and can be taken once a day. In general, it is not sedating and is well tolerated by the majority of people who take it. Some people develop stomach pain or nausea; in these cases it is particularly important to take the drug with meals. A few people lose weight. There is a slight risk of kidney stones, and a few people develop decreased sweating, which can cause overheating.

Pregabalin (Lyrica)

Only one drug is likely to enter the U.S. market in the next few years: pregabalin (already approved in some European countries). This was developed as an analogue of gabapentin, one of the safest drugs available and one with relatively few side effects. Its mechanism in the body seems similar. Pregabalin promises to have these same advantages, but it is approximately six times more powerful in controlling seizure as is gabapentin. It also does not have dose-dependent absorption, which limits the amount of gabapentin that a person can absorb in a single dose. Pregabalin has been studied extensively for anxiety and neuropathic pain and will likely be approved for these uses as well. It also seems to have sleep-stabilizing properties and may improve sleep in some people with epilepsy.

Another, More Complicated Case: Part Two

Manisha and her doctor discussed the various other drugs available and decided to try zonisamide. When she first took zonisamide, a relatively low dose of 200 milligrams per day, she was having about two complex partial seizures a week, and many smaller ones every day. With the medication, her seizures became confined to the night. This was understandably much less disruptive for her. As her other drugs were stopped and the amount of zonisamide was increased, her seizures improved to where she had maybe one per month, all during sleep. Over several months, she became noticeably more confident. She said that she finally felt her life was moving forward. She now believed that she would make something of her life.

Manisha felt that zonisamide caused a tremendous improvement in her life, as she never had to worry about seizures happening in public or in dangerous situations such as while driving. In fact, zonisamide enabled her to live her life in a way she felt was completely normal: the goal of all epilepsy treatments.

While no drug works in all people, there are similar success stories with all of the new agents. In some whose seizures were controlled but who suffered severe side effects, the new medications allow them to feel normal all of the time. In others, like Manisha, the new drug has virtually or completely eliminated seizures, giving them a new outlook and freedom in their lives.

Choosing the Right Drug

As you can see from the discussion above, a large number of drugs are available for epilepsy. Most people have partial seizures, for which most drugs are effective, but a number of choices exist for each of the other seizure types. How do you and your doctor choose the right medication for you?

The first goal of treatment is, of course, complete control of seizures. Despite extensive research over many decades, however, there is no concrete evidence that any of the drugs for a particular seizure type is more effective than another. Doctors may have opinions, sometimes based on considerable experience, but no major differences have been proven. This may seem confusing since all of these drugs have different actions in the brain. Probably, just as there are many ways a seizure can begin, there are also a number of ways that it can be stopped. "Tweaking" one group of neurochemicals can therefore be as effective overall as a completely different effect on another neurochemical pathway.

Besides effectiveness, other differences can be important. As we've explored, some drugs (particularly phenobarbital, but to a lesser extent phenytoin and carbamazepine) are more likely than others to be sedating. These drugs may be more likely to cause problems with thinking and memory than some of the newer agents. Some drugs (phenobarbital, phenytoin, carbamazepine, and valproate) may interfere with the body's normal production of hormones. In most cases this does not cause an obvious problem but, over time, could be related to infertility (particularly in women) and bone loss, or osteoporosis. These same drugs (valproate to a lesser extent) have a number of interactions with each other and with other prescription and nonprescription medications. These may not be the best choice if you are taking medicines for other conditions.

Phenobarbital, phenytoin, carbamazepine, oxcarbazepine, and topiramate can interfere with oral contraceptives ("the pill"), and accidental pregnancy because of contraceptive failure is more common. A few drugs are more convenient in that they can sometimes be taken once a day; these include phenytoin, lamotrigine, phenobarbital, valproate (in its extended-release formulation), and zonisamide.

Many epilepsy drugs have the potential for rare but serious side effects. Felbamate is probably the most problematic, even though the chance of a severe reaction is still quite small. All anticonvulsants also have the potential for less severe, but bothersome, effects. Mostly these have to do with thinking and alertness and depend on the dosage used. Dizziness and unsteadiness can occur, as can nausea and stomach pain. Many of these effects will disappear after a person has been on the drug for a few days or weeks. Valproate and gabapentin can cause weight gain, while topiramate and zonisamide sometimes cause weight loss.

The main point is that drug choice should be individual. A drug must be chosen that is effective for your seizures and does not cause daily problems. Side effect profile is also individual; a drug that tends to make you lose weight might be beneficial in one person, but is a problem in one who is already underweight. Your doctor should also understand your full medical history to ensure that the drug chosen does not worsen another condition or interfere with other medications you may take. In some condition such as migraine, the correct epilepsy drug may actually help both illnesses. Although serious problems can occur with most epilepsy drugs, they are all very safe. Your doctor should work with you in choosing an initial drug and helping you to understand what to expect. If you think that the drug might be causing a problem, you should discuss it with your doctor and see if there is a different way of taking it, or if the problem may be unrelated to the drug. If the first drug tried is unsatisfactory due to adverse effects or continued seizures, alternatives should be considered (realizing that the next drug could cause different problems, but could also perform better or have no troublesome effects).

Finally, although a study of a hundred people may show no difference in the effectiveness of a drug, in individuals there can be differences. We simply do not yet know enough about epilepsy to understand why one person will respond to a certain antiseizure drug and another will not. Some people may be born with differences in their nervous systems, such that their GABA system is a little less active than another's. These people might never develop epilepsy unless they have the sort of injury described earlier, but then a drug that enhances GABA works better for them, while others may benefit more from drugs that affect sodium channels. Overall, these drugs are rather crude. To correct an error in a small group of neurons located in one part of the brain, the drugs change the workings of neurons throughout the entire brain, sometimes in many ways. It is therefore no surprise that the drugs have side effects. The newer drugs, in general, cause less sedation. With the exception of felbamate, they also interact with other drugs far less frequently, or not at all. This can be particularly important in people who take a number of other drugs, or in women who are taking oral contraceptives. Perhaps the most important aspect of the new drugs is that there are now far more choices, making it easier to individualize treatments. With so many drugs available, it should now be possible to fully control seizures without adverse effects in most people, and this should always be the ultimate goal of treatment.

The following table summarizes information presented above about the most commonly used drugs for epilepsy.

TABLE 8-1: DRUGS FOR EPILEPSY

Drug	Year Approved	Usual Total Adult Dose (mg/day)	Doses per Day	Seizures Treated	Most Common Side Effects	Rare Serious Side Effects
Phenobarbital	1912	90–180	1	Partial	Drowsiness, dizziness, blurred vision, rash	Anaphylaxis
Phenytoin (Dilantin)	1938	300–400	1–2	Partial	Drowsiness, gum thickening, dizziness, blurred vision, hirsutism, rash	Anaphylaxis, aplastic anemia
Ethosuximide (Zarontin)	1960	750–1,500	2–3	Absence	Drowsiness	
Carbamazepine (Tegretol, Carbatrol)	1974	600–1,400	2–3	Partial	Drowsiness, dizziness, blurred vision, rash	Anaphylaxis, aplastic anemia

Drug	Year Approved	Usual Total Adult Dose (mg/day)	Doses per Day	Seizures Treated	Most Common Side Effects	Rare Serious Side Effects
Valproate (Depakote, Depakene)	1978	1,000–3,000	2–3	All	Nausea, drowsiness, weight gain	Liver failure
Felbamate (Felbatol)	1993	2,400–3,600	2–3	All	Anorexia, insomnia, nausea	Aplastic anemia, liver failure
Gabapentin (Neurontin)	1993	1,800–3,600	3–4	Partial	Drowsiness	None
Lamotrigine (Lamictal)	1994	100–500	1–2	All	Rash	Anaphylaxis
Topiramate (Topamax)	1996	100–400	2–3	All	Drowsiness, mental dulling	Glaucoma

Drug	Year	Dose (mg)	Doses/day	Seizure type	Common side effects	Serious
Tiagabine (Gabitril)	1997	32–56	3–4	Partial	Drowsiness, nausea	None
Levetiracetam (Keppra)	1999	1,000–3,000	2–3	All	Drowsiness	None
Oxcarbazepine (Trileptal)	2000	600–2,400	2–3	Partial	Drowsiness, dizziness, rash	None
Zonisamide (Zonegran)	2000	100–400	1–2	All	Drowsiness, nausea	Kidney stones

9

A Chance to Cure: Is Surgery a Possibility for Me?

It was crisp outside, fifty degrees, a bright and burnished October day. Perfect football weather, thought Frank Cicone as he drove the tree-lined road with its curving cathedral of yellow and red, toward Middletown High School. The team had been doing well so far that season; with luck they might make it to the Pennsylvania state finals this time. Andrew had done particularly well. His six-two, 240-pound frame was the anchor in the lineup. Frank tried not to become overly preoccupied with Andrew's health, but it was particularly difficult not to scream in anguish when his son was hit hard on the field. Over the past several games, with no incident, Frank was beginning to relax.

Frank's concern, though, was warranted. For years, Andrew had been having "spells," where he would blank out for twenty seconds or so. At first they joked about his "spaciness," but as time went on, the spells became difficult to ignore. During the longer ones, Andrew would hum to himself, and sometimes his right arm would curl up involuntarily. Afterward, he was often confused for ten or fifteen minutes. Finally his parents took him to their family physician, who immediately diagnosed epilepsy and began treatment with carbamazepine. The spells improved, and two weeks ago the doctor had recommended that Andrew stop medication altogether. He and his family were thrilled. The illness was over.

Andrew actually already knew otherwise. Over those two weeks since he had stopped the drug, three spells had occurred, but one happened when he was alone, and the other two he was able to cover up. Even that Saturday morning, as he prepared for the game, the mild staring spells began again, with their strange confusion; he felt as if he had left this world and entered another. Unlike in the past, however, the spells today were coming one after another. As he was showering, he found himself crouched in the tub, the water still running, not knowing what had happened. As he was dressing, he became entangled in his shirt and could not remember even picking it up to put on. When he arrived at school, an intense nausea rose in him; he ran outside and vomited. He became aware of a hand on his shoulder.

"You okay to play, man?"

"Sure." He blinked. "Yeah, bad fish last night."

The march onto the field revived him. The crowd was cheering, and it felt great to be there. Agility drills. His line raced forward to hit the other; Andrew stopped and found himself standing alone.

"You sure you're okay?" he heard someone say. He began to nod, then blacked out. Somewhere high in his head neurons were spitting, a storm was spreading, and for the first time, Andrew's whole body clenched—not just his arm, not just a fixed stare—but his whole body pitched to the ground, his limbs slamming into the turf, again and again.

Frank had parked in the lot just behind the stadium and had walked up the metal stairway from behind the stands. As he neared the top, the boisterous noises of the high school crowd suddenly gave way to a worried murmur. His step quickened; although he prayed he would be wrong, he somehow knew he wasn't. He raced up the last step and looked down on the mud-splattered field. He saw a small circle of players and managers crowded at the forty-yard line. He could not see who was in the center but it didn't matter: he knew. As he raced down toward the field, the coach noticed him and nodded, his face tense. The other players parted, and Frank saw his son motionless on the field, breathing heavily, unconscious.

Just as his father arrived the convulsion had stopped, but Andrew would be confused and disoriented for another quarter of an hour. Years later, after Andrew's epilepsy had been cured, Frank confessed to me that he had had that feeling of dread each and every time he had gone to see his son play ball even though he had only had a grand mal seizure that one October afternoon.

Epilepsy cured? The neurosurgeon where I trained in the subspecialty of epilepsy had a favorite question that he would repeat at every possible opportunity: "Name a chronic neurological disease that can be cured." In my training as a neurologist, I learned that most of what we did was not curative. Cure happened in the emergency room, when the chicken bone was plucked from the throat . . . in the operating suite, when the cancer was excised . . . on the medical ward, when the infection was diagnosed and the correct antibiotic administered. Unfortunately, most of the illnesses neurologists treated could not be cured. A stroke victim might recover, but the damage was done and our most important job was prevention of another. Parkinson's disease, Alzheimer's disease, Lou Gehrig's disease—all have treatments that may slow the process and improve the symptoms, but the disease is there for life.

Epilepsy is different. Usually, treatment with drugs is enough to stop all symptoms. In relatively mild cases, which fortunately make up the majority, the brain can even heal itself with time, restoring itself to rightful balance. But about 30 percent of people do not have their seizures completely controlled with medication. For some of these more intractable cases, those who have failed to have complete control with the first, the second, or the tenth drug or combination of drugs, the disease can be cured with surgery. Although surgery can be enormously successful, with up to 90 percent chance of cure for the best candidates, it is grossly underutilized. Many primary care physicians, and even neurologists, view surgery only as a "last resort" when all other options are exhausted. This hesitancy can have serious consequences for the drug-resistant person. Decades of continuous seizures cause lifelong patterns of fear and dependency that no surgeon's scalpel can undo.

Andrew was one of those people, and he taught me what it really means to be cured. I met him when he was a twenty-one-year-old college student. His mother had called, reporting that his seizures had increased from a few a month to many every week. Although they were minor, she wanted to know if there were better treatments. His neurologist, a colleague of mine, had just moved out of state, so Andrew and his parents made an appointment to see me.

Andrew's seizures had begun with five episodes before the age of one, all when he had a high fever. Febrile seizures during infancy are common; up to 5 percent of the population has had this kind of convulsion, and typically there are no long-term complications. In retrospect, though, Andrew's were different: he had several, and after one of them he was temporarily paralyzed. We suspect that these "complicated" febrile seizures make the later development of epilepsy more likely. Sure enough, at age nine Andrew began to have complex partial seizures. They always began with a vague feeling that something was going to happen, a sense of pressing immanency. Like many people, he could never really describe the feeling, but once said he sensed things crawling inside his body. If someone asked him something, words sounded "like a foreign language." He could only mumble "uh-huh." He drooled and his right arm curled. Afterward, he felt exhausted. On four occasions Andrew's complex-partial seizures spread outward, igniting his entire brain: grand mal. That's what had happened to him on the football field.

By the time I saw Andrew, he had already been prescribed several standard medicines: carbamazepine, phenytoin, and valproate, all of which helped but did not heal. His life was far from normal. Andrew did well in school because of his diligence and determination: he worked hard even when he was tired from seizures; he kept his body in top physical condition with constant exercise despite feeling weakened by medication. At one point, he was taking thirty-six pills a day and still seizing.

For Andrew's parents, there was constant worry. Every time the phone rang, they imagined him in the hospital, his mouth full of blood and his limbs scraped raw, or worse. They always picked up the receiver with a sense of dread, followed quickly by keen relief: he was okay. Andrew's parents loved to watch their son play football, loved to see him leap and run on the field, his body, sometimes so solid in their eyes, other times teetering on an invisible edge.

I first saw Andrew with his parents, who did most of the talking. Andrew was tall, well built, and good-looking with sandy blond hair and clear blue eyes. He was shy, didn't really look at me much, and answered only when I addressed a question directly to him. Answers came mainly from his mother, a pleasant woman who clearly cared deeply for her son.

As mentioned, Andrew was similar to many people in being unable to describe the feeling he gets before a seizure. "It's just like something's about to happen. Like a wave goes over me." Sometimes a good description can help pinpoint where in the brain the seizure is originating, so it's worth pursuing. "Does it ever feel like déjà vu? Like the events around you have happened before?" This sensation is common in temporal lobe epilepsy, an area closely involved with emotion. Other people may feel sudden intense fear or ecstasy (Dostoyevsky described that feeling as "total harmony"). Andrew couldn't tell me anything more definite, but I knew his seizures had to spread to the posterior temporal lobe, because in an epileptic episode he was incapable of understanding speech, an activity linked to that lobe. We know that speech is usually located in the dominant hemisphere, on the opposite side of the dominant hand. For right-handed people, over 95 percent of the time, speech is located in the left cerebral hemisphere. For left-handed people it's more complicated: it can be in the left, the right, or both hemispheres, with roughly equal frequency. Because Andrew was right-handed, I could be reasonably sure that his seizures came from the left temporal lobe of his brain.

Andrew had already tried three medicines. Although sometimes the fourth or fifth will work, the chances diminish dramatically as the trials extend. After two drugs have failed to work, the chances of a third completely stopping the seizures is less than 10 percent, and the odds continue to worsen with further drug trials. Andrew, of course, wanted to be seizure-free, and that's what really matters. Imagine: you're meeting with a client, about to settle a deal that will make your firm a lot of money, or you're on a first date with a woman you've found attractive for months and finally got the nerve to ask out. But you know that, at any moment, you might curl up, drool, or blankly stare. Even if it only happened once a year, you'd never know when a seizure would strike. Such a life is difficult, undermining confidence at every turn.

Andrew ended up trying another drug, a new one at the time, called gabapentin. His seizures decreased from daily to about three a week. But it was still three a week—and who knew where and with whom. So we began the long evaluation for surgery: a frequently frustrating process that involves trying to pinpoint the renegade tissue, and evaluate it for an excision that will not disturb its neighbors: speech and movement centers, locales of love and judgment. This evaluation is performed systematically, almost always in a comprehensive epilepsy center. These centers were established mainly in the 1980s throughout the United States, and they represent a significant advance in the treatment of complicated epilepsy. Epilepsy centers are staffed by a group of professionals trained in different specialties—neurology, neurosurgery, neuropsychology, psychiatry, radiology, nursing, and social work—all of which are necessary for the treatment of complicated cases and all of which are essential for the complete evaluation of candidates for epilepsy surgery. Sometimes these are termed phase 1 and phase 2 evaluations. In phase 1, tests are performed to prove that the person has epilepsy, and to attempt to determine precisely where it originates in the brain and whether the affected brain can safely be operated on without causing a problem. All of these tests are described in more detail in chapter 7. Phase 1 testing consists of the routine EEG and MRI (which almost always have already been performed); video-EEG (a prolonged EEG with simultaneous video) to record actual seizures; neuropsychological testing; and a Wada test to evaluate memory and language. Additional images of the brain, such as SPECT or PET, are sometimes performed. All of the tests in phase 1 carry little or no risk to the person. Sometimes, phase 1 testing gives enough information to direct surgery. If surgery seems like a reasonable possibility, but the area for operation is not precisely found or seems close to an essential brain area, a phase 2 evaluation may be recommended. This consists mainly of an operation to implant electrodes directly onto the brain followed by more video-EEG monitoring. With these electrodes, it is usually possible to precisely determine where the seizures begin. Brain mapping is usually performed as a part of this: the electrodes are stimulated from the outside, which can give a precise map of the brain's functions (language, movement, and so on) and how they relate to the site of seizure onset.

Types of Surgery for Epilepsy

There are four types of surgery for epilepsy: localized resection (or removal of brain tissue); corpus callosotomy; hemispherectomy; and multiple subpial transection (MST). Another surgical procedure, implantation of a vagus nerve stimulator, is not actually brain surgery, as the device is implanted in the neck. Localized resection is performed far more often than any of the other brain surgeries and has the greatest overall chance of seizure freedom. It is not, however, possible in all people with epilepsy.

Localized Resection

The bulk of this chapter concerns this technique, as it is by far the most common procedure. A localized resection requires finding the focus, or site of onset, of the seizures and removing it surgically. This is not an option for people in whom the focus cannot be found, where several foci exist, or where removal of the focus would also cause some serious problem, such as inability to speak or to move a part of the body.

In most people, however, the seizures arise from a fairly well-defined area of abnormal brain. Removing it will therefore stop the seizures. Although we know that particularly with long-standing epilepsy, wider areas of the brain may be irritable and perhaps capable of producing seizures, nevertheless, removal of the focus (and if possible, other highly irritable areas) has a high chance of giving a person freedom from seizures. This is the outcome roughly 80 to 90 percent of the time if the seizures begin in the temporal lobe, and 50 to 60 percent if they begin elsewhere. Many people who do not become seizure-free will have substantial improvement.

Corpus Callosotomy

The corpus callosum is a large bundle of nerve fibers that connects the two sides (or hemispheres) of the brain. It is the major pathway for the transfer of information from one hemisphere of the brain to the other. For example, if you pick up a hot coal with your left hand, the sensation will be interpreted in the sensory cortex of the right parietal lobe. If you now want to describe it and you are right-handed, your main language centers are probably in the left hemisphere, meaning that the information from the right parietal cortex will flash across the corpus callosum to the left hemisphere so you can exclaim, "It's hot."

The corpus callosum is also a major pathway for the spread of seizures. It is particularly important in certain generalized seizures, particularly atonic seizures, but also generalized tonic-clonic seizures. In people with generalized seizures, remember that there is no discrete focus; the entire brain seems irritable. Cutting the major pathway of spread, the corpus callosum, is unlikely to completely stop these seizures but can make them much smaller. In people who suffer frequent injuries due to falling during seizures, these smaller seizures are much less likely to result in harm.

In other people, a discrete focus cannot be found. There may be one, but the seizure spreads so fast that it seems to start on both sides of the brain. A corpus callosotomy in these people can reduce the severity of seizures and can subsequently make it possible to find the seizure focus, allowing another, more definitive surgery.

Some people wonder how they will function with all the connections between the two sides cut. Literally, it would seem, the right hand would not know what the left hand is doing. In fact, in most cases, a complete corpus callosotomy is not performed. About four-fifths is cut, leaving the posterior part intact. This may result in temporary difficulties, particularly with speaking, but these rapidly resolve. A number of other connections between the hemispheres can be used to transfer information, the largest of which is called the anterior commissure. These remain intact, so the corpus callosotomy is unlikely to cause permanent problems and can benefit people by reducing the severity of seizures and, in a few cases, by eliminating seizures.

Hemispherectomy

Hemispherectomy is the removal of an entire hemisphere of the brain. The brain is not actually removed entirely, but parts that are not are disconnected from the other hemisphere. This operation was found to cause fewer complications than the actual removal of such a large amount of brain tissue.

Why would such a dramatic operation be necessary? In a rare condition called Rasmussen's encephalitis, one hemisphere degenerates. Seizures are frequent and typically do not respond well to anticonvulsant medications. As this disease progresses, the functions of this half of the brain, including movement, feeling, and (if the dominant hemisphere) speech are affected. Ultimately, the person can become hemiparetic, or unable to move half of her body normally (like a person who has had a large stroke). Seizures can continue and cause further disability. In such a person, there may be little further problem in performing a hemispherectomy, as the hemisphere is already severely affected. The operation can stop the seizures and the further disability they cause. The remaining hemisphere can then work to its full potential. When the procedure is performed in young children (younger than about age five), the brain is still "plastic" and the remaining hemisphere can take over most of the functions of the diseased one. Even in the best cases, though, movement on one side will never be normal.

Multiple Subpial Transection

Multiple subpial transection (MST) does not remove brain tissue. Rather, the surgeon makes numerous shallow cuts along the surface of the brain. This leaves the neurons at the surface intact, but cuts some of the connections among them. In theory, this can limit the ability of a seizure to spread. Because brain tissue is not destroyed, MSTs can be performed even when the seizure focus is directly atop "eloquent" cortex, such as language or motor areas, without permanently damaging these functions as a resection could.

There is still controversy over how well this technique works for seizure control. In many cases, a seizure focus is found that partly overlaps eloquent cortex. Brain tissue is resected where possible, and MSTs are performed over the remaining, eloquent areas. In these cases it is impossible to know whether seizure improvement was due to the resection, to the MSTs, or both. Most epilepsy specialists believe that a resection of a seizure focus is more likely to result in seizure freedom. If this is not possible, however, MSTs are a good alternative that may also result in significant improvement, or even freedom from seizures.

Vagus Nerve Stimulator

Unlike the operations described above, the vagus nerve stimulator is a device that is surgically implanted into the body for seizure control. Technically, it works somewhat like a pacemaker in the heart. The surgeon places a coil of wire around the vagus nerve, one of many nerves in the neck. The left nerve is always used, because the right vagus nerve helps control heart rhythms, and stimulation with the device could cause cardiac complications. The surgeon tunnels the wire under the skin and connects it to a flat device (a few square inches in size) that is placed in the area of the shoulder, usually under the muscle. The device contains a battery and a small computer that can be programmed from outside the body, using a sort of "wand."

The vagus nerve stimulator is then programmed to give intermittent pulses of electricity to the vagus nerve. Usually the stimulus lasts for about thirty seconds and is given every five minutes. The programmed duration and intensity of the stimulus are usually increased gradually, about every two weeks, allowing the person to become acclimated. Tingling, changes in voice, coughing, and swallowing difficulties can occur while the vagus nerve stimulator is activated. Usually this occurs for a short time after the stimulus is increased and dissipates over the next few days. Some people will continue to have minor effects whenever the stimulus is on, particularly a slight hoarsening of the voice.

It is not actually known how the vagus nerve stimulator works to improve seizures. In theory, the stimulus travels up the nerve into the base of the brain (nerves can conduct impulses in both directions). The stimulating impulse goes into the brainstem, where the "nucleus" of nerve cell bodies that make up the vagus nerve reside. These neurons will then "backfire" and send impulses to other brainstem structures to which they are connected. In this way, conduction in the brainstem is altered and, theoretically, thereby improves seizure control. The improvement does not seem to occur just when the device is stimulating, but through ongoing changes in the electrical systems of the brainstem.

People with the vagus nerve stimulator can also carry a magnet, which they can use to trigger the stimulator immediately (in addition to its programmed cycle). The theory here is that if a person knows when a seizure is starting, indirect stimulation of the brainstem using the vagus nerve stimulator can prevent the seizure from spreading. This is not useful for the many individuals who do not experience warning signs and does not work in all who do. A few, however, can stop their seizures in this way.

In clinical trials of people with hard-to-control seizures, the chances of improving seizures with this device are similar to those of trying a new medication. Therefore, once several drugs have failed to fully control seizures, the chance that the vagus nerve stimulator will completely stop them is small (less than 10 percent). It can, however, result in improvement. Unlike with drugs, there is no delivery of chemicals to the rest of the brain or to the body. The only possible side effects (besides those of the operation itself) are cough, swallowing difficulty, or changes in the voice, usually only during the time the device is stimulating. Initial evaluation for the vagus nerve stimulator should be similar to that for epilepsy surgery, as the best surgical candidates have a much greater chance of cure with surgery than with the vagus nerve stimulator.

Phase 1 for Andrew

I had obtained detailed descriptions of Andrew's seizures through my discussions with him and his parents, but now I needed to see them firsthand. We also needed to record Andrew's brain waves with an electroencephalogram (an EEG) during a seizure. The only way to accomplish these goals was to admit Andrew to the hospital and to wait for him to have an epileptic episode while we recorded videotape and an EEG. We admitted him and reduced his medications to speed things up. This is typically the first step in the evaluation of epilepsy after the first two or three medications fail to completely stop the seizures. It will establish, in most cases, roughly where the seizures originate and is therefore critical in determining the potential risks and benefits of surgery. Sometimes this step shows that the episodes are not epilepsy at all, but another condition (described in chapter 6); this can be true even when an experienced neurologist thinks it is epilepsy.

In all, we recorded four of Andrew's seizures. We saw, on tape, his inability to understand a nurse's simple commands. We saw he could not lift a cup of water or wave his hand. A look of dense blankness filled his eyes. Three of the seizures stopped here, but one seizure worsened, and we saw Andrew have a grand mal. As the seizure ignited the motor cortex, both arms and legs contracted with immense force and he gave a piercing cry.

The EEG showed exactly what I suspected. Just as Andrew pressed the alarm to indicate he was going to have a seizure, the electrodes over the left temporal lobe began to pick up a different signal. That area began firing all of its neurons at once, and the happy disarray of his normal brain gave way to spikes of an angry discharge firing six times per second. The discharge spread to the rest of the left side, then to the whole brain as the convulsion started. We know that if a generalized convulsion lasts over an hour, neurons can begin to die. But Andrew's stopped in less than a minute, with no outside help. We quickly put him back on a full dose of medicine and discharged him from the hospital. I prayed he wouldn't have a seizure during his interview with Chase Manhattan Bank the following week.

I was, at that point, pretty certain that Andrew had temporal lobe epilepsy. That is good, because we know that this type has the highest chance of cure with surgery: between 80 and 90 percent. But we weren't finished. He had already had an MRI, a detailed image of the brain obtained by magnetic waves passed through the head. If the MRI identifies an abnormality—a cluster of neurons where they should not be, a scar from a decades-old injury—that can help to pinpoint where the seizures arise *if the finding correlates with the information from the EEG*. Andrew's MRI was repeatedly normal, which made things a bit more complicated.

He needed neuropsychological testing, typically the next step in a surgical evaluation. Everyone has strengths and weaknesses—some are better at math, others at learning complicated procedures, still others at putting words together. In cases of epilepsy, we may find that the area of the seizures doesn't work as well. For example, if the left temporal lobe (an area usually associated with language) has been affected by seizures, the person may not do as well on verbal testing. In Andrew's case, he did do slightly worse on the verbal tests than on those that did not involve language.

He next needed the Wada test, described in chapter 7. Andrew came back to the hospital for this test, in which we inserted a catheter into the femoral artery at his hip, feeding it up the artery and into his brain. He was then injected with sodium amytal, a barbiturate, which caused him to safely have something like a stroke for about five minutes: one side of the body cannot move. Under these conditions, we can evaluate just how the other side works. If the right brain is knocked out, but the person still can speak, then speech is on the left side. Memory testing is also critical: in people like Andrew, the affected temporal lobe will often support short-term memory poorly or not at all.

When Andrew's left internal carotid artery was injected with anesthetic, his right side was quickly paralyzed. He couldn't speak, indicating that his speech was located in the left hemisphere. He didn't close his eyes when asked. He was quite awake, though, and looked at the eight objects and pictures shown to him. As the anesthetic wore off, he started speaking, but strangely. When asked to say "limes are sour" he said "simes are flower." He remembered all of the objects shown to him earlier.

With the opposite injection, his right brain was inactivated. His speech was a little slurred, because his left face was paralyzed, but he talked fine and followed commands. He could name everything he was shown. But when he recovered, he could only remember half of the things he was shown. So, as we suspected, the memory in his left brain (the side still working during this part of the test) was not as good as that in his right brain. It's impossible to know whether this is a cause or an effect of seizures, but it is common. Maybe whatever caused the seizures in the first place also damages memory. Or maybe seizures almost every day, although small, take a toll on brain cells so stressed and sensitized. In any case, there was now even more evidence that Andrew's left temporal lobe was the problem, and since it was already damaged, he had less chance of a problem if it was removed.

Whenever someone with epilepsy is being considered for surgery, most centers present the person at a multidisciplinary conference attended by neurologists/epileptologists, neurosurgeons, neuropsychologists, nurses, and social workers involved or potentially involved in the person's care. That is because the decision to operate is never completely clear. If a person has appendicitis, for example, the need for surgery is almost always clear-cut because without it there is a high chance of serious illness or death, yet no potential problems arise in living a full life without an appendix. On the other hand, epilepsy may in one case be severely debilitating or even life-threatening, while in another it may have no effect whatsoever on a person's quality of life. In addition, removing part of the brain always carries the risk of some neurological problem. These two issues especially must be weighed before a recommendation is made. Sometimes, the decision is relatively clear either way. A person who sometimes sees flashing colors for a few seconds and nothing else may have no real benefit from surgery to stop these, even if they happen every day, yet the surgery has a strong possibility of adversely affecting vision. Surgery in such cases would rarely if ever be recommended. Another person who has grand mal seizures every day with a clear source shown by testing that could safely be removed would almost always be recommended for surgery.

Andrew was like most people: he fell somewhere in between. While his seizures were frequent, he was able to function quite well. They were reduced with medicine. The chance of affecting his memory with surgery was real but small. But his life was clearly affected by the seizures. He could not drive. He was at constant risk of injury. And although he was able to finish college and had a decent job, he was performing below his full potential. Socially, he remained withdrawn. So we had little doubt that his life would be improved if his seizures were stopped. But what part of his brain would need to be be removed, and at what cost to him?

Phase 2 for Andrew

Sometimes, with the results from the video-EEG, the MRI, the neuropsychological testing, and the Wada, we have enough information to be comfortable about where the seizures come from and whether that area can safely be removed. In Andrew's case, although most of the information pointed toward the left temporal region, his MRI (which often shows subtle abnormalities in people like him) looked completely normal. We knew the area, but not the precise location. Because we also knew that his language areas were not too far from the source of the seizures, we needed more information prior to proceeding with surgery. So Andrew needed the most definitive test: a recording of seizures directly from the surface of his brain. For this, he needed to first have an operation where electrodes were implanted into his brain in the suspicious region. The skull was removed above his left ear, and electrodes were placed on the rumpled surface of the temporal lobe. Other electrodes were pierced directly into the brain with a stiff wire, to reach the inner area called the hippocampus. The hippocampus lies deep within each temporal lobe and is not easily reached with surface electrodes.

Having a patient go through this is always frightening. Here was a young, highly functional man whom I had recommended for brain surgery. Were his seizures really bad enough to warrant it? Would he be one of the few (about 1 percent) with serious complications? What if we were wrong, and the electrodes were not covering the correct spot?

Andrew went through the initial surgery with no problem. He went back to the Epilepsy Monitoring Unit and waited for seizures, as before. He had the first one four days later, showing a crystal-clear seventeen-hertz spiking pattern as his hippocampal neurons began the seizure. He had several more, and they were all identical. Within a week, he was taken back to the operating room to have the electrodes removed and the originating area of his seizures taken out.

Even with the sensitivity of intracranial recording, not every person has seizures that are this clear. Sometimes, the start may be so subtle that several seizures must be analyzed to see the pattern. Some people have seizures that start in more than one location; in those cases, it must be decided whether more than one spot can be operated on, and if not, which is the most troublesome area. Rarely, despite these recordings, the exact spot of seizure onset cannot be seen, or it is found but is atop an essential area (such as speech or movement) that would cause more harm than good to remove. Overall, though, the vast majority of people can be helped by surgery.

Seeing the living brain still evokes the same sense of awe I experienced the first time I saw the brain of a live person during a neurosurgical rotation in medical school. When the triangular wedge of Andrew's skull was again removed, I could see the plastic grid of electrodes on its surface. Now we removed the grid, and according to the recommendation of everyone involved in Andrew's care, Doug, our epilepsy surgeon, prepared to excise a piece of gray matter. It is always so strange to do this or to see this done: chiseling off a fleck of what makes a person who they are. We almost never observe obvious personality changes with this type of operation, but is that because we don't really know what to test? Is it possible that we will cut out the words to an old tune or forever disrupt the circuit Andrew used to remember the first crush he had in high school?

Doug lifted the electrodes, and I looked at the gentle pulsations of Andrew's left temporal lobe. Over the next few hours, Doug would remove the tip of this temporal lobe, a piece of tissue about the size of a plum. This would allow him to access the real problem: the hippocampus, like a small asparagus spear, which he would completely remove. To me this was the most frightening part: the hippocampus is the seat of short-term memory. In a famous case of early surgical treatment of epilepsy, a man referred to as H.M. had both hippocampi removed for intractable seizures. He was cured of his epilepsy, but his short-term memory was totally destroyed. He recalled everything from his childhood, but could not recall the names of the doctors and nurses who treated him every day. Because memories must first be held in the hippocampus before permanent storage, H.M. was also unable to create new memories. As years passed, the only memories he had became more and more outdated. He began to mistake his own photograph for that of his father. H.M. was, and remains, frozen in time.

Most people can function quite well with only one hippocampus; now we never remove a hippocampus without ensuring that the opposite one functions. Andrew's tests showed that his right hippocampus was fine. His left did not work well, but still had some function. Removing it was unlikely to render him another H.M., but he could have difficulty particularly with word memory, since this is preferentially on the left. We could only hope that stopping the seizures would have a favorable enough impact on memory to offset the loss of one ailing hippocampus.

Doug sutured the dura (the tough outermost coving of the brain), replaced the piece of Andrew's skull he had removed, and closed the scalp. The affected area would rapidly fill with cerebrospinal fluid. The remaining brain might scar a bit at the edges, but would not grow back.

Regardless of the ultimate surgical recommendation (focal resection, corpus callosotomy, hemispherectomy, MSTs, or the vagus nerve stimulator), the evaluation of most people is fairly similar to what Andrew went through. First, a neurologist needs to determine that the person is not responding adequately to medication. While there are various opinions on this, most epilepsy specialists agree that, if seizures continue after adequate trials of two different seizure medications, further investigation is warranted. The person should already have received routine testing, including an EEG and an MRI. Occasionally an abnormality will be seen (such as an arteriovenous malformation) that makes surgery a good early option. More often, video-EEG monitoring is performed to investigate the seizure type, make sure that it isn't something other than epilepsy (found in about 25 percent of refractory "epilepsy" cases), and get a preliminary idea of whether surgery would be an option. Besides the testing Andrew underwent, many people receive other types of brain imaging. SPECT (single-photon-emission computed tomography) uses a radioactive isotope to take a snapshot of blood flow in the brain. Images can be either interictal (where the seizure focus shows reduced flow) or ictal (where it shows increased flow due to increased activity during the seizure). Comparison of the two types can give a clear picture of the onset site. PET (positron-emission tomography) is of several types, but the most common measures the metabolism of glucose, which can be decreased at the seizure focus. Some groups are beginning to use functional MRI, which can localize various functions (including language and perhaps seizure onset).

I have seen incredible results with all of the surgical techniques, but I have also seen my share of disappointments. Temporal lobe resection is the most rewarding, as it has the best odds for success. Ultimately, though, despite all of the complicated evaluations, surgical treatment of epilepsy is still a bit crude. We do our best to narrow down an area of dysfunction, then cut it out, no doubt removing a certain amount of normal brain, which, one hopes, will not be missed. There will no doubt be more elegant treatments in the future. For now, though crude, surgery does work in the majority of carefully chosen cases.

The decision to pursue or undergo epilepsy surgery is always individual. Whenever seizures are not completely controlled after trying two or more drugs, and the seizures interfere with the person's life, surgery should at least be considered. At the least, a high-quality MRI and video-EEG recording should be performed. These will show, first of all, whether the condition is actually epilepsy and assure that the correct types of drugs are being tried. If the diagnosis is not epilepsy, more effective treatment can then be prescribed. These tests will also give a preliminary idea of the possibility of success with surgery: if the seizures originate in the temporal lobe, the chances of surgical success are quite high and might push the physician to recommend further testing toward surgery. If the site of seizures is in another location or is unclear, it might make more sense to try further nonsurgical approaches (other drugs or the vagus nerve stimulator), as surgery would be more difficult and less likely to be successful.

The morning following Andrew's surgery, I went to visit him in his hospital room. He was sitting up, fully awake, looking better than I did most mornings. With careful testing, I could not detect even a subtle neurological problem.

Andrew has not had another seizure since then. One month after surgery, he came to my office for evaluation with his parents, as before. I could already detect a change: the shy, reserved youth was more authoritative and confident than I had ever seen him. For the first time, Andrew did most of the talking. By three months he was coming to his appointments alone. He now had a job at Merrill Lynch and was succeeding well at it. While always pleasant and friendly, he became outgoing and funny. His medicines were decreased after a year. I rarely hear from him now because he is busy living his life, just like everyone without epilepsy. While he was fairly successful with seizures, he has truly excelled without them. His mother feels that she and her husband "won the lottery" with the successful surgery, finally able to let go of their constant worry about their son.

When I asked Andrew what life was like without seizures, he grinned. "I don't miss them one bit." He felt as if every single aspect of his life, everything that he does, was positively affected by the surgery. The most important thing for him, though, was that the worry was gone. Before the surgery, he was never without the fear of having a seizure. He said that he always felt he needed to stay in the New York area. He was terrified of being anywhere his parents could not get to. He never traveled abroad, and when he traveled at all, it was with his family. Throughout high school and college, he stayed home when others went on spring break. Now he goes where he wants to. After working for years in international finance, for the first time he was able to consider a job abroad, something unthinkable to him a few years before. He can get on a plane now. He can live where he chooses. How strange that by removing a bit of the world inside his head, we gave him a wider world outside his head, where there is work to be done and people to love. Andrew has a life he can live now, and in his own words, he is in command of it for the first time.

10

Seizures Are Random—or Are They? Day-to-Day Things That Might Affect Seizures

Every person with seizures, no matter how random, tries to find a trigger that causes them. Did it happen because I fought with my husband the day before? Did I have too much wine? Chocolate? Herbal tea? Several of my patients have even attempted to relate seizures to the phase of the moon. Such behavior is natural. Seizures are a largely random event and represent a total lack of control for the person, and it is human nature to look for reason and order. Finding the trigger for the seizures would mean that order and predictability could be returned to their lives.

Specific triggers for seizures do occur in what have been termed stimulus-sensitive epilepsies. The most common of these is writing epilepsy, where seizures occur specifically when manipulating a pen or pencil, but even this is exceedingly rare. Strange, isolated cases do occur as well; in one medically proven case, a woman invariably had a seizure upon hearing the voice of a newscaster, Mary Hart (this was verified with video-EEG monitoring). For the vast majority of people, external factors do have an influence, but never to the extent that seizures can reliably be predicted. The most common of these are alcohol and drugs, hormonal changes (in women), sleep deprivation, and stress. But these are influences, not causes.

Kyle had not had a seizure in seven years when I first saw him. After a long discussion, we agreed to discontinue anticonvulsant drugs. He did well for another two and a half years, then he called because his wife had witnessed him having a generalized tonic-clonic seizure in his sleep.

In this extraordinary case, a specific trigger could be identified. Kyle worked for a charitable foundation headquartered in the World Trade Center. Many of his coworkers were lost on 9/11, and the ensuing weeks were occupied by memorial services intermixed with struggles to keep the organization's various projects running. He felt he could not possibly do enough for the bereaved families, and this combination of extraordinary stressors almost certainly contributed to his isolated seizure. In the end, we decided to take a wait-and-see approach and not to restart treatment. Extraordinary stress can provoke seizures even in totally normal people—as was shown in a study of American soldiers in Vietnam who were evacuated when the United States left Saigon (about one in a thousand otherwise healthy young men experienced a seizure after days of severe stress, sleep deprivation, and often alcohol). Knowing which factors can influence seizures can not only improve the lives of people with epilepsy, but can shed light on what makes the brain work—and become overworked—in all of us.

Factors That Can Commonly Affect Seizures

In general, stress on the brain and body increases the likelihood of seizures in most people with epilepsy. In the case of Kyle, above, the trigger was extraordinary psychological stress. While control of the factors discussed in this chapter does not prevent the need for medical treatment, people with epilepsy should be well aware of these influences on their disease and, as much as possible, adjust their lifestyle to minimize the chances of worsening seizures. On the other hand, your goal should always be to live as normal a life as possible. Stress, sleep deprivation, and other potential influences on seizures are a part of life. As with all people, there must be a balance between avoidance of factors that could worsen seizures and living a full life.

Sleep Deprivation

We know that sleep deprivation places a stress on the body and mind. In people with epilepsy, this can increase the chances of a seizure. It is not known precisely how this occurs, and it is also quite variable. Some people find that even mild sleep deprivation (getting a few hours less than a full night's sleep on a single occasion) usually results in a seizure the following day. This is most common in juvenile myoclonic epilepsy. In most, however, the relationship is a lot less clear. Sometimes the person can get little sleep for several days and not have a problem; another time under similar circumstances a seizure will occur. Often, it is not clear whether the sleep deprivation had anything at all to do with the seizure.

People with epilepsy who are sensitive to sleep deprivation need to pay careful attention to getting enough sleep. They may need specific measures when traveling between time zones to minimize the potential for sleep disruption. One way is to take melatonin (three to five milligrams, available in most drugstores over the counter) at bedtime in the new time zone. The best method is actually to begin adjusting your bedtime by taking melatonin before leaving, moving the dose one hour ahead or behind per day toward the desired bedtime in the new time zone. Short-acting prescription sleeping pills, such as Ambien or Sonata, can also help. If people with epilepsy do become sleep-deprived, they need to be particularly cautious the following day, avoiding driving and if at all possible remaining in a safe, supervised setting, knowing that a seizure could occur.

For most people, however, a little extra attention to sleep is all that is required. Sleep is good for general health as well, and it benefits all of us (in our society of minimal sleep) to allow ourselves the rest our bodies need. Of course there will be nights when a person will get less sleep—when the baby is irritable, or when a major project is due. In most people with epilepsy, occasional mild sleep deprivation is not a problem, and it is sufficient to simply keep this in mind along with other general health issues.

Emotional Stress

Emotional stress is probably the most common factor that people relate to seizures: for example, this one happened because my husband and I were arguing . . . that one because I was upset over balancing the checkbook. In reality, though, the connection is rarely that specific. There were probably many times when a seizure did not occur under similar (or even worse) stress.

It is true, however, that heightened anxiety increases the chances of seizures. I sometimes tell people that if they could live on a tropical island with all of their needs catered to, they would have fewer seizures, but this is rarely possible. Because nobody can remove all stress from his life, it makes more sense to hold yourself to more realistic standards while taking steps to do things that decrease overall stress. Exercise, particularly aerobic exercise, is one of the best ways to release stress. It is important for everyone, but particularly for people with epilepsy. I recommend doing whatever you find most enjoyable: walking, jogging, swimming, tennis, aerobics classes—the type of exercise is not nearly as important as doing it regularly. If you don't enjoy it, you won't do it.

Many people also find meditation and yoga helpful. These are popular in most areas and can easily be learned by taking classes or by watching instructional videotapes. Along with other relaxation techniques, yoga and meditation could improve seizures, but will even more likely improve sleep (perhaps indirectly improving seizures) and overall well-being.

Physical Illnesses

Any kind of illness places a stress on the body, whether a simple cold or a difficult cardiac ailment. Illnesses with fever may be more likely to induce a seizure. It is rare, however, that illness reliably provokes a seizure in anyone.

Because illness can increase the chances of a seizure, it pays to stay in overall good physical condition. This includes regular exercise, attention to sleep habits, and a healthy diet. People should also have regular appointments with a primary care physician even if they are seeing a neurologist regularly for seizures—annually if they are in good health, but more often if another illness requiring treatment is present.

Drugs and Alcohol

Many of my patients ask me about the effects of drugs and alcohol on epilepsy. There are two factors to consider: the effects of these agents themselves, and the potential effects on the medicines you may be taking for epilepsy.

Some drugs are highly epileptogenic and can produce seizures in normal people. The most common of these is cocaine, but other stimulants like amphetamines (speed) or ecstasy are potentially harmful to all, and people with epilepsy should particularly avoid them. Others, like marijuana, can certainly cause health problems but are unlikely to cause an exacerbation of epilepsy.

Alcohol in small quantities is rarely problematic for people with epilepsy. For a few people, even a single drink can be risky, but they usually find that out quickly. People with juvenile myoclonic epilepsy often fall into this category, probably because of the sleep disruption caused by alcohol. Most people don't realize that alcohol disrupts sleep; in fact it is commonly and erroneously used by the public to induce sleep. But in most people, one drink, particularly with a meal, carries no appreciable risk and to some can be important to quality of life. Larger quantities of alcohol can be a problem. Drinking to intoxication probably does increase the risk of seizures. Furthermore, intoxication will significantly interfere with sleep.

What about interactions between drugs and alcohol and your seizure medicine? The problem with most illicit drugs is that these interactions simply are not known. The drugs should be regarded as having the potential to decrease the effectiveness of most anticonvulsant drugs (with the possible exceptions of gabapentin and levetiracetam, which have no known interactions with any drugs). Generally, alcohol will not interfere when taken in moderation, but if used in excess it can change the metabolism of many drugs. People taking sedative anticonvulsants regularly, including phenobarbital, clonazepam, lorazepam, or diazepam, should avoid alcohol altogether because the sedative effects are additive, potentially causing severe drowsiness or even coma.

Factors That Can Affect Seizures in Some People

Flashing Lights

Many people have heard that flashing lights, strobe lights, or the flickering of television or computer screens can cause epileptic seizures. Perhaps the most widely publicized occurrence of this "photosensitive" epilepsy occurred on December 16, 1997, when millions of Japanese children were watching their favorite television show, *Pokémon*. This remarkably popular program was shown by six television stations, one of which (TV Tokyo) had a 16 percent audience share in that time slot. In this particular episode, a flashy cartoon sequence included rapidly flickering red tones at about twelve times per second, lasting for up to four seconds. Unknown no doubt to the producers of this program, this frequency is in the range of highest sensitivity in susceptible individuals. In fact, strobes of this frequency are commonly used in EEG laboratories to cause epileptic discharges. While photosensitive epilepsy is not common, the huge number of children watching this program led to over seven hundred being rushed to hospitals with seizures, the majority of whom had never had one before. Thousands of additional children (about 5 percent of the children watching) had more minor symptoms, including smaller seizures, headache, and nausea.

Photic stimulation, or flashing lights, is one of many phenomena that can result in seizures. This is, however, a very specific type of epilepsy. Only about 5 percent of epilepsy patients (100,000 in the United States) are susceptible to it, but another 30,000 or so are "photosensitive." These people would show a response in an EEG lab, but will never experience a seizure because of flashing lights encountered in daily life. The types of flashing lights most people experience are nowhere near the range that would potentially cause a problem. Strobe lights are usually far too slow, and the flickering of screens much too fast. The *Pokémon* episode was an exception. Having said this, however, one everyday situation where photosensitive people can actually get into trouble is riding in a car down a leafy street. The constant flicker of sunlight shining through the leaves can actually result in seizures. For the vast majority of people, however, flashing lights are not of concern, and susceptibility is routinely tested when an EEG is performed.

Menstrual Cycle

Many normal processes in the body can influence the chance of seizures, and hormonal changes are no exception. In women, two prominent hormones, estrogen and progesterone, have opposite influences on the chance of seizures. Estrogen makes seizures more likely, and progesterone seems to inhibit them. During the monthly cycle of all women of childbearing age, levels of estrogen and progesterone fluctuate (unless a woman is taking an oral contraceptive, which itself contains hormones). In some women with epilepsy, seizures are more likely when estrogen levels are relatively high (midcycle and just before menstruation) and less likely at other times of the month. These hormones occur in men, but at much lower levels so they don't usually influence men's seizures.

It can be tricky to determine whether seizures actually correlate with the menstrual period, and a systematic approach should be used. Both seizure occurrence and the menstrual cycle need to be carefully recorded for at least several months using a calendar. Even better is the use of ovulation predictor kits, available in any drugstore, to determine the time of the month when ovulation occurs: another time of potentially high risk in some women. If seizures do reliably happen only or almost only at a certain point in the monthly cycle, this information could be important for finding the best way to treat them. Medications can be adjusted during high-risk days, or hormonal treatments given in addition to anticonvulsant medication. Hormonal treatment generally consists of an oral contraceptive drug, sometimes in a constant dose rather than the typical three weeks on, one off, used by many women. Oral contraceptive use seldom actually worsens seizures; however, this can happen, particularly if the agent used has a relatively high amount of estrogen. Of course, strategies for hormonal use need to be discussed with your physician.

Individual Triggers

Certain situations or conditions can reliably cause seizures in some people, although this is unusual. Individual triggers vary widely among people but tend to be incredibly specific in an individual. I have had patients who have seizures when listening to Billy Joel's "Piano Man," when reading, and when eating. In a very few cases, the specific activity is the only time a seizure occurs. Many more people suspect that certain things such as exercise or arguing with their spouse bring on seizures, but not all the time. As in looking for a relationship between a woman's period and seizures, sometimes one has to be very scientific to sort this out, carefully recording each seizure and each time the specific activity (such as exercise) occurs.

How seizures become linked to a certain activity is not known, but is fascinating. Most likely, the area of the brain that is associated with the activity (the music area for the person who seizes on hearing Billy Joel) is close to that which is abnormal. Hearing the song activates the abnormal brain area, which then causes the seizure. In many people, even when this happens, it can fade over time as (hopefully) the seizures get better and as the connections in the brain change.

In discussion with my patients, I find it is helpful to think graphically of the many things that can influence whether a seizure occurs (Figure 10-1). The top panel shows a person who does *not* have epilepsy. As each day goes by, the body is subjected to external factors, some of which make seizures more likely (sleep deprivation, certain drugs); others make them less likely. These influences are constantly fluctuating, bringing a person closer to or further from the seizure threshold. In a person without epilepsy, the threshold will *usually* never be reached (although it is possible—remember that isolated seizures occur in up to 10 percent of people without epilepsy, or they can occur because of an infection, drugs, or other factors known or unknown).

The second panel shows a person with epilepsy. The same factors occur from day to day, but in this person the seizure threshold is lower, so that seizures can occur more easily. The goal of drug treatment is to raise that threshold, ideally back to that of a person without epilepsy. Paying attention to the factors discussed in this chapter may also help to stay below the seizure threshold.

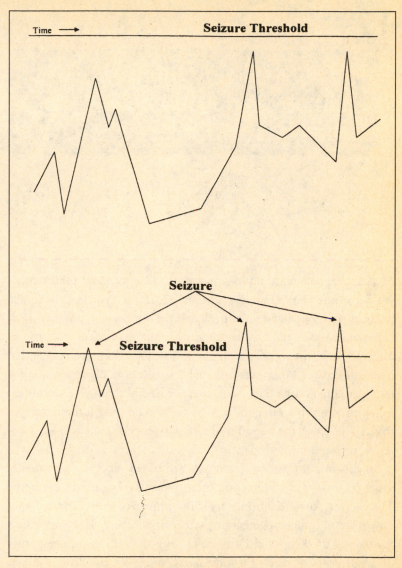

FIGURE 10-1: Graphic Depiction of Factors That Influence Seizures

11

Thinking Outside the Box: Alternative and Herbal Treatments for Epilepsy

Many people with any disease would like to avoid taking medications. They find the idea of taking a synthetically made medication foreign and unnatural, and (in the case of epilepsy) medications generally need to be taken daily for many years. Additionally, persons with epilepsy may be taking alternative or herbal preparations for other reasons, such as to relieve depression or for memory enhancement, unaware that the herbal treatment is interfering with their anticonvulsant, potentially rendering it ineffective or toxic. Therefore, the question of alternative treatments comes up often in my sessions with patients.

In general, alternative treatments are based (at least to some degree) on rational thoughts and facts about the brain and the body. They differ, however, from medically approved treatments in that they have not undergone careful scientific scrutiny. To be approved for use by the Food and Drug Administration (FDA), a drug must undergo years of careful testing. First, safety of the drug is tested typically in normal individuals, followed by testing in the group of people for whom it is intended. Next, effectiveness is proven using one of several rigorous scientific study designs. The most common involves testing effectiveness in a "double-blind, placebo-controlled" trial. *Double blind* means that neither the doctor nor the person knows exactly what the person is receiving, so that unbiased, objective determinations of effectiveness can be made. A placebo-controlled trial tests a drug against a placebo or "dummy pill," which has no effectiveness.

It may seem that persons with seizures could easily tell if their seizures had gotten better, and that these involved and expensive studies would therefore not be necessary. In reality, though, it is difficult to be certain that any improvement (or deterioration) is directly due to the medicine itself. Seizures are random; therefore, some variation could be due to chance. There is also a very real "placebo effect" with seizures as with most diseases. In the placebo effect, persons who believe they are receiving a drug—but in fact are not—will have about a 10 to 30 percent improvement in seizures. The reasons are not known. It may be that people in trials get better medical care for the simple reason that they are followed more carefully. They may also be more scrupulous about their overall care—reliably taking other medications, getting enough sleep, and exercising properly. Finally, there may also be an effect simply due to positive thinking. We know that the mind has the capacity to improve real, physiological parameters such as blood pressure and the immune system, so it may well be that the placebo effect results in real physiological changes that have a positive impact on disease.

Why many of the treatments below may work is not well understood, since they are not studied as carefully. They may not work at all, but the belief that they work may give some improvement. If the treatment works, does it really matter why? Finally, however, some alternative treatments can be harmful. This is the most important aspect to be careful about. Even "natural" substances can have toxic effects (poisonous mushrooms, poison ivy) and can even be fatal. These agents are drugs, just like any medication obtained by prescription. Anyone who is considering an alternative treatment should first make sure it will not be harmful in itself or interact badly with their anticonvulsant drugs or with any other medication they may be taking or any other condition they may have. The Epilepsy Foundation believes that alternative therapies are acceptable as long as the person continues with traditional therapies, and as long as the alternative and traditional therapies do not conflict.

Dietary Issues and the Ketogenic Diet

Many people become excited when they learn of the "ketogenic diet"—developed decades ago as an effective treatment for epilepsy. The ketogenic diet works on the principle that the brain (unlike any other organ) is capable of running on only two fuels: glucose (a direct product of sugar and carbohydrate metabolism) or ketone bodies (formed by the metabolism of fat). The latter normally comes into play only when a person is starving. The ketogenic diet, through severe restriction of carbohydrates, limits the availability of glucose to the brain and tricks it into using ketone bodies for energy. The majority of calories in this diet are obtained from fat rather than carbohydrates, with additional calories from protein.

Using this diet to control epilepsy may sound more natural than taking a medication, but in fact it is quite the opposite and can be dangerous. Creating this artificial state of high fat intake and apparent starvation causes a number of other problems, including probable increased risk of atherosclerosis and kidney stones. Practically, it is extremely difficult to maintain, as every ounce of food must carefully be accounted for to maintain the ketotic state while providing essential nutrients. While successful in a small number of people (nearly all children), the ketogenic diet should be undertaken only in special cases. It is most often considered in children who have not responded well to medication and are not surgical candidates. It is easier to administer in children who are unable to eat normally because of severe cognitive or physical problems and receive nutrition through a tube directly placed in the stomach.

My patients often ask me whether other diets can be helpful in epilepsy. None are known, although a small study showed that the Akins diet could improve seizures possibly because of the production of ketones. In any case, proper nutrition is always important. Some anticonvulsant drugs tend to deplete the body of the vitamin folic acid, so that some people (particularly women, as this vitamin is essential in pregnancy) may need to take this as a supplement. Some anticonvulsant drugs can also increase the risk of osteoporosis, so that sufficient calcium should be included in the diet or taken as a supplement. Women are at greater risk for osteoporosis particularly after menopause and should pay particular attention to calcium intake. Proper exercise helps all people with preserving healthy bones as well.

Herbal Treatments

Many herbal and "natural" remedies are readily available over the counter in drugstores, nutrition centers, and even supermarkets. A study published in the *Journal of the American Medical Association* in 1998 estimated that 12 percent of the U.S. population takes at least one herbal medication. The number of people taking prescription drugs was even higher—18 percent. It is therefore important to know the implications of herbal treatments not only in themselves, but also in how they can affect the actions of prescription drugs.

Unlike prescription medications, virtually no FDA regulations govern the sale of these agents. These products are not considered drugs, but dietary supplements and are therefore subject to the much less restrictive guidelines of the 1994 Dietary Supplement and Health Education Act. Manufacturers must ensure that any claims made on the label are true and have evidence to back these up, but there are no specific standards as to what the evidence must be. Theoretically, if a few people took an herbal remedy and got better completely by coincidence, this could be used as evidence that the herb was effective even if hundreds of others did not improve or even worsened. Manufacturers can also claim on the label that a product affects a certain part or function of the body so long as it does not claim to be effective for a specific disease. Sometimes, information is not on the label but is easily available in the form of pamphlets or verbally from staff. The Internet also contains a great deal of information of varying degrees of depth and accuracy on these preparations. For these reasons, claims of effectiveness for herbal products must always be viewed with caution. On the other hand, a few dietary supplements have been studied for treatment of certain diseases in well-designed clinical trials, similar to the standards required by the FDA, and those exceptions are described below.

Dietary supplements have other potential problems compared to FDA-approved drugs. The contents of herbal dietary supplements can be highly variable for a number of reasons. They often contain several potentially active ingredients, and amounts of each can be variable. As the content of these products is not regulated, dosages among different manufacturers and even among batches made by a single manufacturer can vary. The amount on the label does not necessarily correspond to the amount actually contained in the bottle. Unlike prescription drugs, where the dose delivered must be carefully proven, herbal preparations may contain only a fraction of the claimed dose or may be prepared in such a way that the body does not absorb the active ingredient. Finally, other compounds (either naturally occurring or contaminants) may be included that are not listed in the ingredients, increasing the risk of adverse effects or harmful interactions with other drugs.

Below I'll discuss specific dietary supplements that have either been used for epilepsy or are popular overall and are therefore more likely to be used for other reasons in persons with epilepsy.

Naturally Occurring Hormones, Including Melatonin

Hormonal issues can be important in control of epilepsy, particularly in women. Many women find that seizures tend to occur at certain points in the menstrual cycle. In a few, this relationship is precise, such that seizures occur only at ovulation or menses. This makes physiological sense, as the main hormones that fluctuate during a woman's menstrual cycle, estrogen and progesterone, have known antiepileptic (progesterone) and epileptogenic (estrogen) properties. This relationship and potential treatments are discussed in chapter 10.

Another naturally occurring hormone, melatonin, is available commercially in health food stores and drugstores and is widely used as a treatment for mild sleep disorders. In normal people, melatonin is secreted from the pineal gland according to a precise cycle over a twenty-four-hour period. Around bedtime, there is a dramatic increase in melatonin, facilitating sleepiness. The secretion of melatonin then falls during the night, and by morning the levels are low. This cycle is an important part of our circadian rhythm: the system that regulates numerous body functions over a twenty-four-hour cycle, the most obvious of which is the sleep-wake cycle. Some other functions that are part of this include temperature (body temperature is normally lowest in the early morning), and secretion of some other hormones (such as growth hormone and cortisol).

The release of melatonin is sensitive to light, and its release is typically resynchronized every day by daylight and nighttime darkness. When time zones are traversed, daylight "hits" the pineal gland at a different time, initially disturbing the melatonin cycle and contributing to jet lag. Ultimately, the light cues in the new time zone will reregulate the melatonin cycle.

Why is melatonin important in epilepsy? Humans' need for sleep is overlooked in general by Western society, despite the demonstrated importance of sleep in overall health and well-being. People with epilepsy need quality sleep more than anyone else, since severe sleep deprivation can actually cause seizures, and anticonvulsant medications can increase the tendency toward drowsiness. Melatonin supplements may be helpful simply in regulating the normal sleep-wake cycle. Melatonin supplements can also have direct anticonvulsant effects in photosensitive epilepsy and partial epilepsy. Absent large trials, however, the actual usefulness of melatonin to treat seizures independently of its beneficial effects on sleep is unknown. Anecdotally, I have recommended melatonin for some people who may benefit from its sleep-regulating properties. A few have had their seizures improve, which may be due to improved sleep. But even if an anticonvulsant effect is present, it is probably mild and does not change the requirement for anticonvulsant medication.

Kava

Kava has been used for anxiety and insomnia. Several well-designed trials have studied kava's effectiveness against anxiety; some (but not all) showed that it performed better than placebo. Some of the actions of kava overlap with those of some anticonvulsant drugs, including inhibition of calcium and sodium channels and increases in inhibitory (GABA) neuronal transmission. Claims of usefulness in seizures have never been proven in any scientifically sound trial.

The main worrisome effect of kava is toxicity to the liver, which has actually led to its removal from the market in several European countries. Kava can rarely cause disorientation and sleepiness, although some reasonably well-performed trials show no such effects in the majority of people, or effects similar to those occurring with placebo.

Saint-John's-Wort

Saint-John's-wort has widely been used for depression. Several well-designed clinical trials showed that Saint-John's-wort improves mild depression more than placebo, and some studies have suggested it is as effective as a prescription drug. Other studies, particularly studies in people with more severe depression, showed it was ineffective. A guideline issued by the American College of Physicians–American Society of Internal Medicine said that Saint-John's-wort may be considered for mild to moderate depression for short periods.

Most trials with Saint-John's-wort show that the majority of people have no side effects when using it. When side effects are seen, they are most often stomach problems, dizziness, tiredness, dry mouth, headache, allergic reactions, frequent urination, or sexual problems. The main concern for people with epilepsy is that Saint-John's-wort can increase the metabolism of drugs that are broken down by a particular enzyme system in the liver, called CYP3A4. Its use could therefore result in other drugs becoming less effective or ineffective. Taking Saint-John's-wort with a type of antidepressant known as the selective serotonin reuptake inhibitors or SSRIs (such as Prozac and Zoloft) can cause a severe reaction in anyone. Of all the dietary supplements, Saint-John's-wort is the most likely to cause changes in the effectiveness of other drugs, although no specific changes in drugs for epilepsy have been found.

Valerian

Valerian has been used for depression and anxiety, and for insomnia. No well-designed clinical trials have assessed its effectiveness for these conditions. Like certain prescription drugs, valerian can cause tiredness, and long-term use can actually lead to drug dependency. Some information suggests that valerian can interfere with the effectiveness of phenytoin; therefore people who take phenytoin should avoid also using valerian.

Ginkgo Biloba

Ginkgo biloba is used for memory enhancement, particularly in people with Alzheimer's disease or strokes. Some fairly good scientific trials suggest its use results in mild improvement, although the effect is not large. Most people do not have side effects, but headache, stomach problems, diarrhea, and allergic reactions can occur. There are a few reports of seizures with ginkgo biloba, although even if true, this is probably rare. Ginkgo biloba probably does not interfere with epilepsy drugs.

Ginseng

Ginseng is taken for a number of reasons, including memory and mood enhancement, vitality, and improved libido. No well-designed clinical trials have been conducted to assess whether these claims are true. There are no known serious side effects, and ginseng does not usually cause appreciable side effects. It is unlikely to interfere with other drugs, including those for epilepsy.

Echinacea

Echinacea is reported to decrease the duration of the common cold. Trials have been performed, some of which seem to show that it is effective; others show that it is not. There are no known serious side effects, and echinacea is unlikely to interfere with other drugs.

Ayurveda

Ayurveda is a comprehensive system of alternative medicine that originated in India thousands of years ago. It includes herbs, nutrition, exercise, massage, yoga, and other spiritual and physical treatments. Some of its herbal remedies have been proven to decrease the concentration of one anticonvulsant, phenytoin. Using ayurveda herbs with phenytoin could therefore result in seizures due to inadequate levels of phenytoin, even if a constant dose recommended by your doctor was maintained. There are unproven claims that one type of ayurveda medicine, shankhapushpi, improves seizures, but its interaction with prescription seizure drugs probably makes its use more detrimental than advantageous.

Silymarin

Silymarin (milk thistle extract) has been used for liver disease. Although the effect is not proven, it is said to change the metabolism of valproic acid such that levels fall in the bloodstream, potentially making this anticonvulsant ineffective. People taking this drug should therefore avoid this agent.

Hawthorn

Hawthorn extract has been used for heart failure, and some well-designed trials show improvement in people's symptoms. As heart failure is a potentially serious condition, however, hawthorn should be used only under the supervision of a doctor as progress (and need for other treatments) must be carefully evaluated. Side effects produced by hawthorn include stomach problems, chest pain, and dizziness. It does not directly interfere with other drugs, but again, use in treating heart failure (particularly in combination with prescription drugs) needs to be carefully monitored by a physician.

Saw Palmetto

This product has been advocated for the treatment of enlarged prostate. Saw palmetto does change hormone levels, so there is good reason to believe that it could work for this condition. Some carefully designed trials did in fact show improvement in symptoms when people took saw palmetto; others suggest that an effect may be present but is not large enough to be important. Side effects of saw palmetto have usually been seen to be no different from placebo, so most people have no problems with taking it. There are no known interactions with other drugs.

Ephedra

Many products that claim to increase energy contain ephedra. Chemically, it is a mild stimulant similar to some prescription medications. Like many stimulants, ephedra has been reported to increase the risk of seizures; products containing this should therefore be used cautiously. A number of other health concerns with this agent has led to the banning of its use in the United States.

Craniosacral Therapy

Chiropractors commonly recommend craniosacral therapy for a number of ailments, including epilepsy. The theory (for neurological diseases) is that manipulation of the neck and spine can affect the flow of the fluid surrounding the spinal cord and brain (cerebrospinal fluid). Practitioners claim that a "cranial rhythmic impulse can be felt at various positions on the skull, that abnormalities in this impulse can be detected, and that they can be corrected by adjustments in the neck and spine.

Craniosacral therapy is generally not harmful. The one serious potential problem is quite rare, but forceful manipulation of the neck may cause a traumatic injury to the vertebral arteries, which supply blood to the brainstem and the occipital lobes. Appropriate manipulation may produce benefits through relaxation of the muscles of the area, improving headache and back pain. This form of therapy (as with all alternative treatments) can also be harmful if medical treatments for other conditions, including epilepsy, are stopped.

For a number of reasons, this therapy is not likely to be beneficial for epilepsy. From a theoretical standpoint, there is no evidence that changes in cerebrospinal fluid flow can be induced by such manipulations. Even if they were, no evidence suggests that any form of epilepsy is caused or in any way related to flow of cerebrospinal fluid. Secondly, studies of the detection of the "cranial rhythmic impulse" show that different examiners cannot reliably detect it in the same person. Even the same examiner does not find the same result reliably. Finally, the validity of this technique has never been tested in a scientific trial; it is based only on single reports of individual people.

Biofeedback

The principle of biofeedback is relatively simple: training the body to voluntarily control a function that is generally under involuntary control. An easy example is heart rate. Most of us never think about our heart rate, but it is relatively easy to be trained to make it go up or down. As part of the training, you need a constant report of your heart rate, so that you can know when it is increasing. Your pulse may be taken with a pulse oximeter (a device that usually goes over a finger or an ear). When the rate increases by five beats per minute, for example, you may hear a pleasant bell tone indicating success. You may not know exactly *how* you got your pulse to increase, but you try to get that bell to ring again. Soon, you can reliably increase your pulse at will.

Biofeedback has been recommended for epilepsy, although its effectiveness has not been well studied. Mostly EEG biofeedback has been used. This technique uses an EEG to monitor changes in certain brain-wave rhythms. EEG biofeedback teaches people to maximize rhythms associated with relaxation. This may help to reduce stress and enhance well-being. So, while not rigorously studied, EEG biofeedback might be helpful.

Acupuncture

Acupuncture is an ancient technique that uses needles to stimulate areas of the body, thereby treating a variety of diseases. It affects the body (and theoretically improves seizures) by several possible mechanisms. Acupuncture may alter electromagnetic fields, activate certain neurotransmitter systems (mainly opiates), and change release of other neurotransmitters or of hormones that act in the brain.

In animal studies of this technique in epilepsy, most show no improvement, with the exception of one study in a small number of dogs. In people, one well-designed trial showed no benefit with acupuncture. Like many therapies in this section, however, acupuncture is unlikely to be harmful unless it takes the place of other, more accepted treatments. A few of my patients have tried this technique, none with any definite success.

Yoga, Meditation, and Relaxation Techniques

Stress apparently plays a role in epilepsy. Many people with epilepsy report that their seizures increase or occur exclusively during times of physical or (particularly) emotional stress. This is a difficult hypothesis to test scientifically, however, because stress is hard to measure objectively. Situations that might be stressful to one person (such as a disagreement with a spouse) might not be stressful at all to another. One study looked at the occurrence of seizures in Israel during the 1980s when random bombings were common. This type of situation, when people know they have no control over whether they might be harmed, is known to be stressful. Surprisingly, though, no increase in seizures was seen in this study compared with the same subjects in more peaceful times.

This is not to say that stress has no role in the development of seizures; in fact, as mentioned, most doctors strongly believe that it does. However, keep in mind that stress is rarely the only causative factor in seizures. It is also impossible to remove stress from your life, so you must find ways to reduce its harmful effects on your body.

Yoga and meditation are relaxation techniques; others include massage and aromatherapy. The goal of these is to reduce the effects on your body of harmful everyday influences or unusual stressors. Reducing stress has many potential benefits. It may improve your ability to fight infection, and it decreases depression. For someone with epilepsy, finding ways of better dealing with stress could result in better seizure control as well. I encourage all of my patients to learn stress-relief techniques, but particularly those in whom stress seems to play a particular role in seizure occurrence.

Alternative medicine is increasingly used in addition to traditional medicine for treatment of many diseases, and certainly for epilepsy. As mentioned, yoga and relaxation techniques probably have the greatest chance of benefit, and virtually no risk of worsening epilepsy. Herbal treatments (used for epilepsy or for other conditions) probably have limited or no effectiveness, with the possible exception of Saint-John's-wort for depression. All of these should be used cautiously due to the potential for interaction with prescription drugs (with the exception of gabapentin and levetiracetam, which are probably unaffected by any other agent, prescription or otherwise). In all cases, use of these agents should be discussed with your physician. Many physicians will not be familiar with alternative medications but should be able to give at least some general guidance in their use. Even those rare physicians who are well trained in the use of alternative medicines will not know every possible interaction with the anticonvulsant you may be taking, but can advise you on the risks and benefits of their use and on finding more information.

Summary: Herbal Treatments That Can Affect Anticonvulsant Drugs

Echinacea   Mugwort
Garlic   Saint-John's-wort
Milk thistle

Anticonvulsant Drugs Potentially Affected

Carbamazepine (Tegretol,   Phenobarbital
  Carbatrol)   Phenytoin (Dilantin)
Clonazepam (Klonopin)   Primidone (Mysoline)
Ethosuximide (Zarontin)   Tiagabine (Gabitril)
Felbamate (Felbatol)   Topiramate (Topamax)
Lamotrigine (Lamictal)   Valproic acid (Depakote, Depakene)
Oxcarbazepine (Trileptal)   Zonisamide (Zonegran)

12

What Happens Next: Future Directions in Epilepsy and Its Treatment

Although it may seem that a lot is known about the origin and treatment of epilepsy, much more is still to be learned. We know that head trauma can result in epilepsy later in life, but we have no way of predicting who will get it. Perhaps more important, if we understood precisely what changes ultimately resulted in epilepsy, drugs or other interventions could be used to prevent it from ever starting.

Additionally, vast numbers of people with epilepsy are still not fully controlled with drugs and are not candidates for surgery. Especially for these people, new treatments are being explored. Many new drugs are being researched that are different in structure and action even from the diverse agents already on the market. It may become possible to inject drugs in high concentrations directly into the brain region affected, without affecting function elsewhere in the brain. Brain "defibrillators" are being researched. These, just as existing devices detect abnormal heart rhythms and deliver a shock to correct them, will do the same for seizures. New surgeries may become more common, including the gamma knife—which uses radiation rather than actual cutting to remove the seizure focus. Finally, new understanding of exactly what causes epilepsy is always refining our use of existing treatments. For those who are not fully controlled today, there is great hope for seizure-free lives in the near future.

New Drugs

Drug treatment for epilepsy fortunately works in a majority of cases, but the perfect drug still does not exist despite many new agents marketed in the past decade. Drugs used today work in many different ways, always with the goal of calming the irritability of the brain without affecting other processes such as alertness and memory. All are, however, rather crude in their interactions with the brain, and side effects are therefore common even when the seizures are fully controlled.

Many drugs are in early stages of development. Drug researchers are looking at new mechanisms of achieving the goal of seizure freedom without adverse effects. Some drugs will affect excitatory systems in the brain—a mechanism not utilized by many of the drugs currently on the market. Several such drugs in initial trials show promising results. Other drugs are modifications of compounds already used in treating epilepsy, with the goal of better control and/or fewer side effects. It is hoped that drugs may become more selective; that is, work only on the damaged systems in the brain, without affecting others. Achieving this may be extremely complicated because the affected system could be different in people who appear to have similar seizures. As our understanding of the mechanisms producing epilepsy increases, our ability to design and use more specific and more effective drugs should increase.

Genetics could play a role in intractable epilepsy. Specifically, epilepsy drugs may not work well in some people for genetic reasons. Maybe their neurotransmitter receptors are a little bit different, for example, such that they function fairly normally in health, but do not react to epilepsy drugs as well when epilepsy is present. If this is true, and if reliable methods for testing people for this are found, these individuals might be recommended for alternative treatments (such as surgery) earlier. It might also be possible to design new drug treatments once this is understood.

Finally, there is great interest in developing drugs that could prevent epilepsy from starting in the first place or, alternatively, from becoming difficult to fully control. Such an advance is problematic because epilepsy undoubtedly develops for a number of reasons, and we do not know precisely how epilepsy results from any of them. Research will probably center on conditions known to have a high chance of giving rise to epilepsy, such as severe head trauma or stroke. Success in this area would be even better than developing a perfect drug for epilepsy: prevention of the disease from ever starting.

New Drug Delivery Systems

Drugs for epilepsy are currently taken mainly by mouth. While convenient, this means that the drug travels not only throughout the brain (including to areas where it may not be needed), but also throughout the body. This means that the drug can act in other areas of the body, such as the stomach (causing nausea), the heart (with a rare potential for arrhythmias), or the bone marrow (rarely causing aplastic anemia).

Mechanisms are being developed for direct delivery of drugs to affected areas. Much of this research is being done for cancer treatment, as people with cancer require toxic chemicals to kill their tumors yet not affect normal areas of the body. These same systems may someday be able to deliver drugs directly to the area of a seizure, potentially making them more effective while virtually eliminating side effects.

Brain Stimulators

Electrical stimulation of certain areas in the brainstem can reduce seizures. This is another technique used more widely in the treatment of other diseases—for example, against certain kinds of tremors, and for severe Parkinson's disease. In epilepsy, its use has been limited, as which area is best for stimulation has not been determined. In fact, it may be different in different people. When used, deep brain stimulation can sometimes reduce seizures, but it has rarely been found to eliminate them. Further work in this area is clearly needed before this could become a commonly used technique. The vagus nerve stimulator, an indirect means of stimulating the brain, is currently used but has limited effectiveness. Variations of this technique might yield better treatments in the future.

A newer, and potentially more promising, technique may be able to stop seizures by delivering an electrical shock directly to the place the seizure starts. This technique is based on the same principle as heart defibrillators, which have been in use for years. The heart (like the brain) works through electrical currents. For the heart to pump effectively, these currents must cycle through a precise sequence with each beat, and they must repeat time and again. A cardiac defibrillator constantly measures the electrical activity of the heart. If the activity becomes abnormal, the defibrillator delivers a shock to reset the rhythm and allow the heart to again pump normally.

The rhythms of the brain are more complicated than those of the heart. It is already possible, however, for computers to detect abnormal brain rhythms associated with a seizure, via electrodes placed directly on the brain, as we've discussed in previous chapters. Preliminary work shows that delivering an electrical shock early in a seizure can stop it or prevent it from becoming larger. Within the next few years, people with epilepsy may have a small device implanted that can detect and stop seizures, probably even before the person knows one is about to happen. This will include electrodes implanted in the area of seizure origin, similar to those now used to record seizures before surgery. The electrodes will be attached to a tiny computer attached to the skull and powered by a small battery implanted in the chest similar to those now used in pacemakers and vagus nerve stimulators. The activity of the device—including how often it is activated—will be monitored in a doctor's office by a remote sensor that can "interrogate" the device from outside the body. By the same external monitor, the doctor will also be able to change the way the device is programmed to maximize seizure control. This system is now in early clinical trials and will likely first be used in people who are not the best candidates for surgery. If this device works well, it could eventually replace surgery for many people whose seizures cannot be controlled with drugs.

Magnetic Stimulation

Transcranial magnetic stimulation is an experimental technique that uses magnetic, rather than electrical, currents to potentially diagnose and treat epilepsy. The result, however, is equivalent to the delivery of an electrical current to the brain's surface, through magnets placed outside the skull instead of electrodes. Magnetic fields can induce an electrical current at a nearby site, even through the skull and brain coverings. In theory, such magnetic devices could stimulate the brain similarly to electronic ones, but without the implantation of electrodes.

There is no well-accepted use for magnetic stimulation at this time. Some epilepsy centers use it to help identify areas of the brain that could be the focus of seizures. Some speculate that the induction of currents through magnetic stimulation could result in improvement in seizures—perhaps similar to the vagus nerve stimulator. But this interesting technique requires much more study.

Surgical Techniques

Technological advances have improved surgery for many conditions, including epilepsy. A surgeon can now view a three-dimensional image of the person's brain on a screen in the operating room, showing precisely where the surgeon is operating inside the brain. Linking this with images showing a tumor or other abnormality can help to safely remove precise areas that may not be distinguishable by sight.

Gamma knife surgery uses precisely focused radiation to destroy areas of the brain without ever cutting the skin. This technique is sometimes mistakenly called laser surgery, but no laser is involved. Gamma knife surgery has been used for people with temporal lobe epilepsy, as well as some other types of epilepsy. People can go home immediately after the procedure. It takes several months for the cells in the area to die, and therefore for seizures to stop. Swelling of the brain can happen, causing headaches, but medication can reduce this effect. Gamma knife surgery is still considered experimental, but as more is learned about the methods, this type of "bloodless" surgery could replace many conventional operations.

PART THREE

Living Well with Epilepsy

13

Staying Safe

Accidents can happen to anyone, and people with or without epilepsy can fall, burn themselves, or have a car crash. In people with epileptic seizures, however, the risks of these types of accidents, as well as a few others, are greater. Obviously the best way to prevent seizure-related accidents is to stop seizures altogether. Fortunately that is possible for most people with epilepsy, but even when seizures continue, most can live normal lives. All people with epilepsy need to pay a little more attention to avoiding situations that could place them at high risk of accidents. In this chapter, we'll explore those risks as well as strategies for avoiding injuries due to them.

Seizures in themselves, with rare exceptions, do not cause injury to the brain or the body. Even with grand mal convulsions, the seizure will resolve with nothing more than muscle soreness and sometimes a bite on the tongue or lip from forceful muscle contractions. When and where a seizure occurs, however, can determine whether a major injury occurs. Even a minor seizure can cause severe injury if one is driving an automobile, where even a brief distraction can result in an accident. For people whose seizures carry absolutely no alteration of consciousness (simple partial seizures), such risks should be minimal or nonexistent. It's essential to understand what is meant by *alteration of consciousness*. Many people insist that they do not "pass out" during a seizure, and they may even think they are fully conscious, but for a time (typically about a minute) they are not fully aware of what is happening. Even that brief alteration of consciousness may pose a hazard. Overall, however, the more severe the seizure type, the greater the likelihood of injury.

Seizures and Minor Injuries

During seizures, even small ones, a person may not be fully aware of what is happening. Typically this is not a problem: he will likely continue to sit or stand (if the seizure is fairly small) for the minute or so before returning to full consciousness. But at times such fleeting lapses could result in injury. For example, if a seizure occurs while a person is cooking or showering, causing momentary unresponsiveness to heat, burns can occur, occasionally severe ones. Injury can happen while working with tools, especially power tools, or people using knives can inadvertently cut themselves.

Does this mean that people with epilepsy should never use tools or cook? We all live with some risk in our daily lives. Obviously, though, for those who have frequent or severe seizures, it makes sense to minimize risk whenever possible. Use irons that have automatic turnoff switches. Try not to cook with an open flame, and use a microwave oven when possible. Showering is preferable to bathing, as there is no risk of drowning. People with children have special risks (see chapter 15).

Seizures and High-Risk Situations

For anyone with seizures, some situations are high risk and should be avoided if at all possible. Driving is a special issue and is discussed below. Many other high-risk situations are optional for most people and therefore easy to avoid. These include sports like skydiving, hang gliding, and scuba diving. Swimming is not particularly risky unless seizures are frequent; however, someone else should be present who knows that a seizure might occur and who is knowledgeable about lifesaving techniques. Team sports (football, basketball, baseball, etc.) generally carry no appreciable additional risk as many others are present who would note the seizure and be able to help in avoiding injury. Common sense is the best guide: if the activity would place the person at serious risk if a seizure occurred, it is best to avoid that situation.

Seizures and Driving

Except in a few large cities with good public transportation, driving is an important part of life and virtually essential for freedom of movement. There are two aspects to driving and epilepsy: the law and the person's health. Any potential for seizure with alteration of consciousness makes driving risky, and for that reason every state restricts driving licenses for people with epilepsy. The specifics of the laws vary by state, but the intent is the same: to prevent people at high risk of seizures from driving. This may seem harsh, but the potential for serious harm if a seizure occurs while driving is great: to the driver, to passengers, to occupants of other vehicles, and to pedestrians nearby.

In a few states (New Jersey, Pennsylvania, California, Delaware,

Nevada, Oregon, and Vermont) your doctor is required by law to report people to the department of motor vehicles if they have had a seizure with alteration of consciousness. Your license may be suspended as a result. In most states, however, reporting is the responsibility of the person with seizures. Typically she will then be required to submit a form completed by her doctor stating the circumstances of her seizures and the date of the most recent seizure. Many states require periodic updating of this information, usually accompanied by a statement from the doctor. Most states stipulate a specified period, from three months to one year, that must pass between the last seizure and reinstatement of the driving license. In several of these states, a general recommendation is made for the time from the last seizure until a license is given, but considerable flexibility is granted for special circumstances. Other states (such as Colorado and Illinois) do not specify a particular period, and the determination is made by the department of motor vehicles on an individual basis, usually with the assistance of a state medical advisory board. In many states, a person with seizures with preserved awareness or seizures that occur only during sleep may drive.

Some people feel such a need to drive that they find ways around the law. In states where the doctor is required to report, they may travel to a neighboring state for care. In states where reporting is the responsibility of the person, they may simply not report it and continue driving. Most of the time, particularly in persons with rare seizures, no problems will occur. However, it only takes one seizure at a bad time—on the highway, in heavy traffic—to result in a serious accident with injury or death to the driver or to other people. If a seizure disorder is found, and the person was told not to drive but did so anyway (or was required to report it and failed to do so), there may be further legal ramifications.

On the other hand, no driver has no risk of accidents, including serious ones. Overall, studies show that persons with epilepsy who do drive (most of them legally) have a lower accident rate than the general population. This may be because they are more careful drivers or because they drive less. On the other hand, studies have shown that among people with continued seizures who drive, 39 percent have seizures while driving, and over a quarter have accidents because of seizures. It is best to follow the laws of the state regarding driving. If special circumstances exist (such as a single seizure when a person was unable to get medication for a few days), this should be explained to the doctor and to the department of motor vehicles in the periodic report and should not result in the suspension of the license.

Besides the law, however, a person must weigh the need to drive against his own safety. If he is at risk for seizures, and someone is riding with him, it may make sense for the other person to drive. If medication needs to be changed, it may be advisable to avoid driving during the transition to be sure that the seizures are controlled on the new medication. In general, the longer the time since the last seizure, the lower the chance of another. *Lower* means just that: while the law may allow a person to drive after being seizure-free for three months, the risk of seizure is low but not zero; at six months the risk is even lower, but still not zero.

A summary of driving restrictions by state is shown in Table 13-1; up to date information can also be obtained from the Epilepsy Foundation's Web site (www.epilepsyfoundation.org).

TABLE 13-1: STATE-BY-STATE DRIVING REGULATIONS FOR PERSONS WITH SEIZURES

State	Seizure-Free Period (months)	Period Flexible?	Frequency of Medical Review (years)	Physician Must Report?
Alabama	6	No	1	No
Alaska	6	No	NS	No
Arizona	3	No	NS	No
Arkansas	12	No	NS	No
California	3,6, or 12	No	NS	Yes
Colorado	None	NA	NS	No
Connecticut	3	Yes	NS	No
Delaware	None	NA	NS	Yes
District of Columbia	12	No	1	No
Florida	24	Yes	NS	No
Georgia	12	No	NS	No
Hawaii	None	NA	NS	No
Idaho	None	NA	1	No
Illinois	None	NA	NS	No
Indiana	None	NA	NS	No
Iowa	6	No	6 mo, then at renewals	No
Kansas	6	No	1	No
Kentucky	3	No	1	No
Louisiana	6	No	NS	No
Maine	3	Yes	NS	No

Maryland	3	No	NS	No
Massachusetts	6	Yes	NS	No
Michigan	6	No	NS	No
Minnesota	6	No	6 mo	No
Mississippi	12	No	NS	No
Missouri	6	No	NS	No
Montana	None	NA	None	No
Nebraska	3	No	None	No
Nevada	3	No	1	Yes
New Hampshire	12	Yes	None	No
New Jersey	12	No	6 mo (for 2 yrs)	Yes
New Mexico	12	Yes	NS	No
New York	12	Yes	NS	No
North Carolina	6–12	No	1	No
North Dakota	6	Yes	1	No
Ohio	None	NA	6 and 12 mo, then each year	No
Oklahoma	12	No	NS	No
Oregon	6	Yes	NS	Yes
Pennsylvania	6	No	NS	Yes
Rhode Island	None	NA	NS	No
South Carolina	6	No	6 mo, then yearly	No
South Dakota	12	Yes	6 mo	No
Tennessee	6	No	NS	No

Continued

State	Seizure-Free Period (months)	Period Flexible?	Frequency of Medical Review (years)	Physician Must Report?
Texas	6	No	1	No
Utah	3	Yes	6 mo	No
Vermont	None	NA	NS	Yes
Virginia	6	No	NS	No
Washington	6	No	NS	No
West Virginia	12	No	NS	No
Wisconsin	3	No	6 mo (for 2 yrs)	No
Wyoming	3	No	1	No

NA: not applicable

NS: not specified

Reference: Krauss et al., "Individual State Driving Restrictions for People with Epilepsy in the US," *Neurology* 57 (2001): 1780–85.

Prolonged or Repetitive Seizures

Most seizures are not, in themselves, dangerous. It is of course possible for a person to bruise or burn herself while she is unaware, and if she falls during the seizure, she could suffer a concussion or a broken bone, but these are unusual. The seizure itself generally lasts at most a minute or two and does not cause any permanent damage to the brain (although the person could be groggy afterward).

Prolonged seizures (known as status epilepticus), however, can be dangerous. The most serious seizure is a prolonged generalized tonic-clonic seizure (grand mal). Permanent disability or death can occur if these last more than an hour. The damage partly comes from associated changes in the body: the constant, violent muscle activity places strain on the heart, muscles, and lungs. The main damage probably results directly from the incredible overactivity in the brain, which exhausts the usual mechanisms for protecting brain cells from damage. If enough brain cells die, problems with movement, speech, or sensation can result. Risks are greatest in older people, as younger brains are more resilient.

With this in mind, any generalized tonic-clonic seizure that lasts longer than about five minutes should be taken seriously. The same is true for repetitive seizures that occur without return to normal between them. If a single or successive seizures occur for five minutes or more, emergency services should be called immediately, so that the person can be taken to a hospital and medication given to completely stop the seizures long before they last an hour (although many will spontaneously stop in the meantime). For persons at risk of prolonged or successive seizures (as in when this has happened before), medications can be given at home to quickly stop these seizures. The most common is diazepam (Valium), which can be given as a suppository (as the person is clearly unable to swallow). It is absorbed this way quickly, almost as fast as if it were given intravenously, and can prevent emergency-room visits.

For smaller seizures, even a prolonged episode is less dangerous, although it should still be taken seriously. There may be time to call the physician for advice. Rectal medication is also effective for these cases and should be considered.

Can You Die During a Seizure?

To the world, she was Flo Jo: the flamboyant track star who appeared in a wild, single-legged bodysuit at the 1988 Olympics in Seoul. She was arguably one of the most successful and recognizable athletes in the world when she retired from competitive running after winning three gold medals in Seoul. But ten years later, she would die in bed from complications of an epileptic seizure. How could this happen? And how can other people with epilepsy avoid it?

Florence Griffith Joyner grew up in the projects, in the Watts section of Los Angeles. She was the seventh of eleven children. Always a bit eccentric, she had a pet boa constrictor named Brandy, whom she bathed and lotioned, saving and painting its shedded skin. She was once sent away from a shopping center because she was wearing the five-foot-long Brandy around her neck. She set high school records in sprinting and long jump, then went to college at California State–Northridge. She dropped out after the first year for financial reasons and worked as a bank teller until she secured financial aid. She also developed a talent for incredibly intricate braiding of hair, masterpieces taking up to thirteen hours to produce.

Flo Jo made her first appearance on the world stage at the 1984 Olympics in Los Angeles. She won the silver medal in the women's 200-meter sprint, second to her teammate Valerie Brisco-Hooks. At the 1987 World Championship in Rome, she also won the silver medal and was part of the winning 100-meter relay team. Her boyfriend, Al Joyner, took gold and they were married a month later.

At the Seoul Olympics, she made her mark. Not only an incredible athlete, she was also a bold performer. Her loud bodysuit, her wildly flowing hair, and her long, painted fingernails (one, of course, painted gold) made her as memorable as did her lightning speed. She left Seoul with gold medals in the 100-meter and the 200-meter as well as the 400-meter women's relay, and a silver in the 1,600-meter women's relay, and she also broke the Olympic record for the 100-meter (10.54 seconds) and the world record for the 200-meter (21.34 seconds). In 1988 she won the Sullivan Award as the best American amateur athlete of the year.

Epilepsy seems to have entered her life after her running success, although the date of her first seizure is not known. She reportedly suffered a generalized convulsion in 1990, and was evaluated in

1993 and 1996 when seizures occurred on airplanes. She reportedly refused further treatment after she was seen at Barnes-Jewish Hospital in Saint Louis.

On September 21, 1998, she seemed tired but otherwise well. She went to bed. Some hours later, her husband found her facedown on the pillow, not breathing. He tried to resuscitate her, but she was dead on arrival at the hospital. Immediately, rumors of steroid and other drug use abounded. The autopsy, however, showed that she had suffered a seizure and, because of her position in the bed, was unable to move to breathe in the sluggish state immediately following, causing her to suffocate in the pillow. A cavernous angioma was found in the left orbitofrontal part of her brain, a usually benign abnormality in blood vessels that can, however, cause seizures. No drugs other than Benadryl and acetaminophen were found in her bloodstream, so presumably she was not taking medication to prevent seizures.

This is a dramatic example, and fortunately unusual. But it is important to remember that particularly during and immediately after generalized tonic-clonic seizures, the body's normal protective mechanisms are compromised. Every effort should therefore be made to eliminate seizures, or at least to limit them as much as possible. Perhaps if Flo Jo had been taking medications to prevent seizures, she would not have died in this way.

What to Do If Someone Is Having a Seizure

The most important thing to do if someone is having a seizure is to prevent her from harm. If she is falling, gently get her to a bed, sofa, or floor. Prevent injury from objects around her. Do *not* attempt to place objects in the person's mouth in a misguided attempt to keep the person from swallowing her tongue. It is actually impossible to swallow the tongue. This practice has unfortunately become commonplace, but actually places both the person having the seizure and the witness at increased risk of injury. The person's teeth may be broken by biting the object, or the object could become lodged in the throat, interfering with breathing. Contractions of the jaw are extremely forceful during a seizure, so the well-meaning onlooker may also find himself with a severe bite to his hand. The tongue is, however, often bitten during a seizure; this cannot be prevented and will, in any case, heal relatively quickly.

If the person remains unconscious after the seizure stops, the safest position for the person is on her side. As secretions (saliva, vomit, blood) are frequently present during seizures, and the person will be unable to protect the airway, this position will allow secretions to drain and prevent the possibility of aspiration of fluids into the lungs. The person may be confused or even violent after a seizure. An onlooker should speak calmly and reassuringly and not try to restrain the person in any way (no holding or pushing), as this can increase agitation. If a person is in a dangerous area (such as a roadway), it is wise to try to lead her gently to a safer place. If law enforcement or emergency medical personnel arrive, onlookers should be sure to tell them that the person has had a seizure. Police may mistake confused behavior for drunkenness or drug use, with subsequent confrontation and restraint of the person, possibly leading to further agitation, violence, and mistaken arrest.

Most people with a known seizure disorder do not need to be taken to an emergency room after a typical, uncomplicated seizure. If injury occurs requiring treatment or if the seizure is prolonged, immediate attention may be required. In other cases, it is sufficient to simply allow the person to recover, review things that may have caused the seizure (did he forget to take medication?), and consider whether the treating physician needs to be called.

14

Other Medical Problems Common in Epilepsy Patients: Depression, Migraine, and Sleep Disorders

People with epilepsy can have other medical problems. Sometimes this is purely coincidental, but some diseases seem to occur more commonly in people who already have epilepsy. Others are closely related to epilepsy, either in terms of treatment or in terms of one disease making the other worse. This chapter looks at specific conditions commonly associated with epilepsy, and how these can affect treatment. The most important are probably depression and migraine, but others such as psychiatric conditions, sleep disorders, and memory problems also commonly occur in people with epilepsy.

Depression

For many reasons, depression is much more common among people with epilepsy than in the general population. This may partly be because of the frustration and helplessness associated with such an unpredictable disease (if seizures are not fully controlled), but the relationship is certainly more complicated than that. The chemical changes in the brain responsible for epilepsy probably overlap those that cause depression. Some studies have actually shown that people who have depression are more likely to later develop epilepsy than people who have never been diagnosed with depression.

All people have periods when they feel sad, and it may be difficult for someone with epilepsy to know whether this feeling is simply associated with having the disease or whether there is an independent (or possibly related) depression that needs to be treated. Symptoms commonly associated with depression include depressed mood, feelings of hopelessness, inability to concentrate, and memory problems. Changes in sleep habits (either insomnia or increased sleep time), changes in eating habits (weight gain or loss), and lack of interest in sex are also common. Most depressed people have some, but not all, of these symptoms.

Recognizing a concurrent depression is important for several reasons. Depression can be a large part of the reason a person is not functioning at his best, even if seizures are completely under control. Recognizing depression should lead to appropriate treatment, which can consist of psychotherapy (talking to a trained psychologist, psychiatrist, or social worker regularly), medication, or both. The most commonly used drugs for depression are known as selective serotonin reuptake inhibitors (SSRIs). These drugs affect a specific system in the brain that relies on the neurotransmitter serotonin. Although seizures are listed as a potential side effect of these medications, this is rarely (if ever) a problem for this class of drugs, and persons with epilepsy can safely take them if needed. Drugs in this class include fluoxetine (Prozac), sertraline (Zoloft), paroxetine (Paxil), escitalopram (Lexapro), and citalopram (Celexa). Another common class of drugs for depression is tricyclic antidepressants, named for their chemical structure. These interact more with the brain systems using acetylcholine as a neurotransmitter. Tricyclic antidepressants are used less often than the SSRIs as they tend to have slightly more side effects, but they can be helpful in some people especially when SSRIs are not. Some examples include amitriptyline (Elavil), nortriptyline (Pamelor), and doxepin (Sinequan). Other drugs include trazodone (Desyrel), which may also be useful for insomnia, nefazodone (Serzone), and bupropion (Wellbutrin). The last drug, buproprion, should be used with caution in persons with epilepsy as it carries a somewhat higher risk of seizure exacerbation.

A few of the drugs for epilepsy can actually also help alleviate depression; while others can potentially worsen depression. This is another important reason why people who think they may be depressed should discuss this with their doctor. In some people, lamotrigine (Lamictal) can help depression. Valproate (Depakote), carbamazepine (Tegretol), and gabapentin (Neurontin) have been used to treat bipolar disease (manic depression), although none of these are particularly good at treating isolated depression. Phenobarbital is probably the epilepsy drug most likely to worsen depression, although this can rarely occur with topiramate (Topamax), levetiracetam (Keppra), or zonisamide (Zonegran). Keep in mind that all epilepsy drugs are designed to work on the brain, so effects on mood, both good and bad, are possible with any of these agents.

Migraine

Migraine is common in people with epilepsy. Sometimes migraine headaches occur completely independently of seizures, but in many cases the occurrence is at least sometimes related to seizures. Occasionally migraine and epilepsy are closely intertwined, and in these people migraine frequently develops after an epileptic seizure. Rarely, a migraine can serve as a trigger for epilepsy. I have seen a few people who do not have independent seizures, but when a severe migraine occurs, it can cause an epileptic seizure. The exact mechanism for this is unclear, but these people probably have an underlying susceptibility to seizures that under normal conditions is insufficient to create an actual seizure. Only when the brain is first affected by the migraine can a seizure occur. In this case, treatment of the migraine is crucial, and independent treatment for seizures may not be required.

Probably more than with depression, careful choice of anticonvulsant medication can help people with migraine. Migraine can be treated symptomatically (acutely, when a migraine occurs) or prophylactically (daily treatment to prevent migraines from starting). Several seizure medications are also effective for migraine prevention even in people without epilepsy, including valproate (FDA-approved for this purpose), lamotrigine, topiramate, gabapentin, and zonisamide. Choice of one of these agents may therefore help with both epilepsy and migraine.

Many medications are used for the acute treatment of a migraine, including nonsteroidal anti-inflammatory agents (or NSAIDs, such as acetaminophen and ibuprofen), other pain relievers (narcotics), and tryptans (such as sumatriptan [Imitrex], frovatriptan [Frova], eletriptan [Relpax], or rizatriptan [Maxalt]). Many agents combine a nonsteroidal anti-inflammatory agent with another pain reliever (typically a narcotic) and/or caffeine (examples of combination agents are Fioricet and Midrin). Caffeine clearly helps when migraines are related to caffeine withdrawal, but also seems to help particularly in combination with pain relievers. The mechanism is unknown but it may help to calm swollen blood vessels. All of these acute migraine medications can be used safely in persons with epilepsy and may be used in addition to medications used to prevent migraine.

Sleep Disorders

Quality sleep is critical to the health of everyone, yet it is often overlooked in our busy society that sometimes views our natural need for sleep as more of an annoyance than a requirement. People with seizures need sleep at least as much as other individuals. Sleep deprivation can worsen seizures, and seizures can cause disruption of sleep. This disruption can make it even more difficult to get proper sleep and can sometimes evolve into a cycle of worsening sleep and worsening seizures. Poor or limited sleep also has clear effects on the daytime abilities of all people, with adverse effects on attention, learning, memory, and the immune system. Decreased abilities to concentrate and remember are already common concerns in people with epilepsy. Poor sleep is also associated with an increased risk of accidents of all kinds, with the most serious being motor vehicle accidents.

In general, people who are sleepy enough that it interferes with daily activities regularly may have a sleep disorder. For instance, anyone who frequently falls asleep when performing quiet activities such as watching television or watching a movie is probably not sleeping enough. Those who fall asleep during more active times— such as in conversation or while driving—probably have a more severe sleep disorder. Excessive sleepiness in people with epilepsy can have many causes, from simply not spending enough time in bed to medications to independent sleep disorders such as obstructive sleep apnea (frequent closing of the airway during sleep, causing suffocation and awakening) or narcolepsy. While sleep deprivation is a stressor thought to worsen epilepsy in many cases, sleep apnea may also worsen seizures because of the severe lack of oxygen it can cause in sleep.

A person with epilepsy who feels unusually sleepy, or who has other symptoms of a sleep disorder (unusual movements during sleep, loud snoring, or particularly periods when breathing seems to stop), should discuss this with her physician. An evaluation by a sleep specialist may be necessary, or a sleep test, called a polysomnogram. A polysomnogram consists of sleeping overnight in a laboratory setting with electrodes attached to measure brain waves, movements, and breathing. Correct diagnosis and treatment of a sleep disorder can improve both daytime alertness and seizures.

Postictal Psychosis

In this unusual condition, someone with epilepsy can temporarily appear much like a person with schizophrenia. Postictal psychosis usually happens after several seizures occur in a short time (typically the same day). The person will initially recover and appear normal. One or two days after the seizures stop, he will begin to act strange. He may become paranoid, believing that the government or family members are plotting against him. He may accuse a spouse of extramarital affairs, stealing, or talking behind his back with no evidence for these accusations. Small things may be blown way out of proportion. Called delusions, these erroneous beliefs may be based loosely on fact (maybe his wife had lunch with a friend, but in his mind the meeting becomes a wild affair). One patient of mine who was having marital difficulties became convinced during a postictal psychosis that his wife was taking his checkbook and going through his desk. This normally quiet person became violent and actually threw a piece of furniture through a window of his home. Another person who was religious believed that God had taken her spirit, and that her body was dead. She therefore refused to eat or to take medications because, she said, "I'm dead. I don't need any of that." Some people in this state even hear voices or have other kinds of hallucinations.

Postictal psychosis always goes away on its own, usually after a week or two. In the meantime, though, the person needs to be carefully watched. Medications, if necessary, are the same ones used to treat schizophrenia but in this case should only be required for a few weeks. Why postictal psychosis occurs is not known, but it must have to do with the process the brain goes through in recovering from many seizures. A few people will have this repeatedly, but it doesn't usually recur even when seizures continue. The best treatment, however, is the prevention of repeated or severe seizures.

Memory Disorders

Many people with epilepsy find that their memory has worsened. They may have trouble remembering where they left their keys or when medical appointments are scheduled. They may find that they are less sharp at work, and tasks that used to be easy take longer or require writing down each step. Names of coworkers or new friends may be more difficult to recall.

Memory is complicated. There are actually many kinds of memory—for names, for reading a map, for putting together a puzzle—and each can be affected differently by epilepsy. Memory also naturally changes as one ages; there may be more difficulty remembering short-term things (where are those keys?) but no problem recalling events from years ago or remembering how to fix a car engine.

The effects of seizures over many years can probably affect memory, though experts disagree on this. It is known that a prolonged grand mal seizure (lasting over an hour) can damage the brain. Small seizures may have no effect whatsoever, even if they occur daily. Larger seizures (complex partial seizures) may cause memory problems if uncontrolled for years, but other people may experience no such problems. Medication for epilepsy can also affect memory, particularly topiramate and phenobarbital.

If deterioration in memory is suspected, the best way to sort out the problem is through careful neuropsychological testing. It is difficult to be objective about your memory (everyone would like it to be better), so an objective test is useful to show if there are problems and, if so, in what types of memory. It can also suggest whether the memory loss is caused by medicines, seizures, or an entirely different problem (such as depression). Proper treatment can then be recommended, be it medication change, counseling, or learning techniques to improve memory.

The main point of this chapter is that treatment of your epilepsy needs to consider treatment of your health in general. Any illness has the capacity to worsen seizures, and any medication may have implications for epilepsy treatment. The conditions discussed above are only a few of the most common examples. Make sure that your doctor knows about any other health condition you have besides epilepsy. Treatment of you as an individual will lead to the best treatment overall.

15

Pregnancy, Childbirth, and Issues Particularly Related to Women with Epilepsy

One of the most troubling aspects of epilepsy, particularly to women, is how it may affect their children. Many medications (including those taken for epilepsy) can increase the risk of birth defects when taken during pregnancy. Parents also worry about whether their disease is inherited. Finally, seizures occurring while one is caring for a small child could place the child at risk if nobody else is present.

The most important fact about parents with epilepsy is that the vast majority of them have normal, healthy children. Most women will need to continue to take anticonvulsant medication during pregnancy, because the risk of seizures during pregnancy (although low) is considered greater than the even lower risk of the medications for birth defects. For example, Joan had her first child while she was taking phenobarbital. Sedation was a constant problem for her, however, so that after her son was born, Joan's medication was changed to carbamazepine. Her problems with sedation persisted, but she was trying to conceive a second child and another change could potentially expose the fetus to two drugs rather than one. She was finally changed to lamotrigine after the third child was born, and had a fourth (a daughter) while taking this drug. Each child was born healthy, but Joan herself lived under the constant threat of a seizure, potentially leaving a baby or young child unattended if only for a few seconds. With a few simple precautions to further minimize this risk (described below), she has had no problems, and fortunately her seizures are under good control. Her story reflects the experiences of the majority of women with epilepsy: some additional anxiety and need for planning, but ultimately similar, mostly happy experiences like those of any parent.

Issues for Women Who Are Not Pregnant

Certain aspects of a woman's metabolism make their treatment for epilepsy slightly different from that for men. The most important aspect (as just mentioned) is probably pregnancy. But epilepsy also has unique effects on hormones in women, and it is often difficult or impossible to separate these from the effects of some drugs on hormones. Epilepsy and epilepsy drugs can cause changes in the menstrual cycle, and also changes in appearance and bone health.

All women with epilepsy should be aware that many anticonvulsant drugs can affect oral contraceptives, potentially making them less effective or ineffective. Sometimes, a pill with a higher dose of hormone can be used, but these may still be less effective (resulting in unplanned pregnancies) or may have associated breakthrough bleeding during the menstrual cycle. Women taking these anticonvulsant dugs should use an alternative method of contraception (condoms, diaphragm, or IUD). Drugs and their effects on oral contraceptives are summarized in Table 15-1.

TABLE 15-1: EFFECTS OF EPILEPSY DRUGS ON ORAL CONTRACEPTIVES

Drug	Effect on Oral Contraceptives
Phenobarbital	Yes
Phenytoin (Dilantin)	Yes
Carbamazepine (Tegretol, Carbatrol)	Yes
Valproate (Depakote, Depakene)	No
Ethosuximide (Zarontin)	No
Acetazolamide (Diamox)	No
Felbamate (Felbatol)	Yes
Gabapentin (Neurontin)	No
Lamotrigine (Lamictal)	No
Topiramate (Topamax)	Yes
Tiagabine (Gabitril)	No
Levetiracetam (Keppra)	No
Oxcarbazepine (Trileptal)	Yes
Zonisamide (Zonegran)	No

Women with epilepsy commonly report that their seizures vary with their monthly cycle. Most commonly, seizures increase at the time of menses or a few days before. In some women, seizures also increase in the middle of the cycle, a time corresponding to ovulation (when an egg is released from the ovary, and the time of greatest fertility). If this pattern occurs, the seizures are called catamenial (relating to the monthly cycle). These seizure increases probably occur due to changes in levels of two important hormones, estrogen and progesterone, that naturally fluctuate during a woman's monthly cycle. Of course, if she is taking an oral contraceptive, these drugs then control the monthly cycle, and there is less variability with most regimens, sometimes improving cycle-related seizures. In fact, placing women on oral contraceptives (or the same hormones in slightly different formulations) has been used as a treatment for catamenial seizures even when this treatment may not be completely effective in preventing pregnancy.

It is often difficult to tell if seizures actually have a catamenial pattern. Seizures generally occur randomly, but it is natural to try to find a pattern. Just thinking back on when they occur can be misleading. The only way to be sure is to keep a log of the days when seizures occur over several months, along with the days of menstruation. Ideally, the day of ovulation should also be determined: this can be done with commercially available ovulation prediction kits, which are easy to use. If seizure are definitely related to the menstrual cycle, one way to improve seizure control might be simple by increasing the dose of medication during high-risk times. An older anticonvulsant, acetazolamide (Diamox), is sometimes used only during high-risk days, in addition to the usual anticonvulsant. As mentioned above, oral contraceptives or other hormonal treatments are also occasionally used for such seizures.

A few epilepsy drugs carry particular concerns for women. Phenytoin has an undesirable cosmetic effect over time, called hirsutism (which means masculinization). This can result in increased hair growth, particularly on the face, and coarsening of facial features. Sometimes this can happen in a matter of months, although some women will not experience it at all, but in general the effect increases with longer exposure to the drug. For this reason, phenytoin is not the best drug for younger women who may need to take it for many years. Valproate has been associated with polycystic ovarian syndrome, a condition characterized by hirsutism, weight gain, and the development of many benign cysts in the ovaries. Many of the older drugs, including phenytoin, valproic acid, carbamazepine, and phenobarbital, result in increased risk of osteoporosis, also a particular concern in women as they are already more prone to this disease than are men. It is not clear whether newer drugs have the same effect. Some physicians recommend both periodic bone-density tests to look for osteoporosis and calcium supplements with or without vitamin D (also important for bone health) in any person at risk for bone disease.

Planning a Pregnancy

Deciding to have a child is a big step for anyone, but a condition like epilepsy adds even more considerations. In the not-too-distant past, women with epilepsy were discouraged, even forbidden, from having children. Although stigma and misperceptions about this issue are now largely resolved, some physicians still mistakenly discourage otherwise healthy women with epilepsy from pursuing pregnancy. Epilepsy experts, however, universally agree that childbirth is a reasonable goal for nearly all women with epilepsy.

Men with epilepsy have fewer potential concerns, as seizures and anticonvulsant use will not directly affect the developing fetus. There are still issues of fertility and heredity, however. With both men and women, concerns should be discussed with the physician as early as possible (preferably before trying to conceive, but otherwise as soon as the pregnancy is recognized), both to minimize certain risks and to understand as fully as possible the implications of epilepsy in parenting. In the case of unplanned pregnancies, the physician should be consulted immediately upon learning of the pregnancy.

Inheritance of Epilepsy

Many people wrongly assume they have a high likelihood of passing on epilepsy to their children. Actually, the risk is low. The genetics of epilepsy is complicated for a number of reasons. First, there are many different kinds of epilepsy, and if inherited, the patterns of inheritance would likely be quite different. Second, the actual biological problem causing epilepsy is not known in the vast majority of cases, making it difficult for researchers to find inheritance patterns. When seizures are caused by a head injury, for example, this is clearly not a condition in which epilepsy would be inherited. Another person might have a similar seizure type but with no history of head trauma. A few types of epilepsy, such as juvenile myoclonic epilepsy, have a higher likelihood of occurrence in children of affected parents, but even in this case the chances are probably 90 percent that the child will not be affected. When the cause of epilepsy is unknown in a person, there *may* be a genetic component, but in general any added risk is 1 or 2 percent if one of the parents has epilepsy (remember that there is a 1 percent risk of epilepsy in general).

Fertility

Fertility is affected by numerous variables (many of which have no relation to epilepsy or its treatment). In general, both men and women have a more difficult time conceiving a child as they become older. The decline in fertility in women is more dramatic, and after menopause women are no longer capable of conceiving a child (unless with the help of techniques such as in vitro fertilization and egg donation).

Some aspects of epilepsy do affect fertility. In women (as mentioned earlier), seizures may alter the normal hormonal changes that happen during the monthly cycle. Since these cycling hormones allow the maturation and release of an egg for fertilization, hormonal disruption from seizures or medications can prevent egg release and, therefore, conception. In men, certain medications (such as valproate) may reduce the ability of sperm to swim effectively, lowering the chance of conception.

As with any couple trying to conceive, it is best to discuss concerns with a gynecologist who is familiar with a person's medical condition. If conception appears to be difficult (usually after six months or so of attempting without success to become pregnant), a fertility specialist should be consulted. If medication is potentially preventing conception, it may be possible to change to a different drug.

Anticonvulsants During Pregnancy: Whether and Which to Take

Considerable research is ongoing on the risks of anticonvulsant drugs in pregnancy, particularly the newer ones. However, much is already known. For the older drugs (phenobarbital, phenytoin, carbamazepine, valproic acid), the increased risk of a major malformation (an obvious deformity, such as heart problems, spina bifida, cleft palate, or a major limb deformity) when a mother takes the drug throughout pregnancy is about 2 to 4 percent. Newer research suggests that the risk with valproate may be higher—6 to 8 percent. These numbers may seem high, but the risk of these problems in the general population (without drugs or epilepsy) is about 1 to 2 percent. The particular problems are about the same for all epilepsy drugs, although valproate carries a slightly higher risk of problems with the neural tube—the structure in the embryo that develops into the spinal cord and brain. Minor abnormalities (a slight change in the appearance of the ears, for example) occur as well.

Some of the newer drugs may carry less risk in pregnancy, but at present this is not known. Several of the new drugs do not cause birth defects in animals and do not produce certain toxins thought to be related to birth defects. Several national and international registries of pregnancies are keeping track of women on each drug who become pregnant to determine the outcomes. In time, this will result in further information, and I suggest that all women taking anticonvulsant drugs participate by calling as soon as possible in pregnancy. It only requires a few phone contacts, and the number is toll-free (888-233-2334). At present, however, the best information suggests that no anticonvulsant drug is definitely better or worse than another for potential birth defects. Therefore, the best drug to take during pregnancy is the drug that is best for the mother overall.

Given that most women with epilepsy will need to continue anticonvulsant drugs during pregnancy, it is important to minimize the risks of drug exposure. The risk from taking two drugs is certainly more than the risk from taking one, so it makes sense to try to control seizures with one drug rather than two or more. A woman who plans to become pregnant and is taking more than one drug should discuss with her doctor whether it makes sense to decrease the number of drugs. As most drug-related effects occur early in pregnancy, the change should be made *before* pregnancy to make a difference. Some doctors also recommend using the lowest dose that is effective for the particular person. In rare cases, it may be possible to stop drugs altogether before pregnancy. For example, it is sometimes recommended that all people should consider stopping their drugs after being seizure-free for two or more years; this should of course be discussed with the doctor first. Also, in rare cases the seizures might be so mild that the mother would not be troubled by them for a few months (during the initial portion of pregnancy). However, in the vast majority of cases the risk of stopping medications (due to potential for injury from seizures) is greater than that of continuing to take them.

Self-Care During Pregnancy

Mostly, a future mother with epilepsy has the same needs as any pregnant woman: maintaining a healthy lifestyle, taking prenatal vitamins, and obtaining proper care by a physician. A few things, however, are different. These mainly have to do with the anticonvulsant drug being taken. A woman's metabolism goes through tremendous changes during pregnancy. Blood volume increases since the placenta needs to be supplied with nutrients. Liver function increases so drugs that are metabolized by the liver last a shorter time in the woman's system than if she were not pregnant. The kidneys also work faster at eliminating drugs. All of these changes can mean that the dosage of drug that was sufficient before pregnancy may not be enough, particularly in later pregnancy.

No matter what the drug, its level in the blood should be known before pregnancy. If not, it should be checked as early in the pregnancy as possible. For most drugs, the level should be checked again at the end of the first and second trimesters, then monthly during the last trimester. In this way, any changes in the level can be corrected (by changing the dose) before seizures occur or worsen. Of course, dosage changes may be necessary at other times if seizures worsen or if side effects develop. The dose will need to be adjusted again after delivery. Breast-feeding maintains elevated metabolism to some extent, so women who nurse may still need a somewhat higher dose than before pregnancy. Women who do not breast-feed can usually return to the dose they took before pregnancy within a week or two after delivery and let their doctor know if adverse effects occur.

Seizures can change, for better or worse, during pregnancy. This probably has to do with hormonal changes, as some women's seizures are sensitive to these, while others are not. Changes during pregnancy are more likely if a clear pattern related to the menstrual cycle is seen before pregnancy. About a third of women find that seizures improve or stop during pregnancy, while in about another third they worsen. Worsening may, in many cases, be due to changes in drug levels as described above; if levels are closely monitored, worsening of seizures is less likely.

It is generally recommended that all pregnant women take prenatally formulated vitamins. Prenatal vitamins differ from other vitamins in, for instance, their higher content of folic acid. If a woman is taking one of the older drugs (phenobarbital, phenytoin, carbamazepine, or valproate), most epilepsy specialists recommend that she take an even higher dose of folic acid—up to four milligrams per day. This is particularly true of valproate, because of its known higher propensity for problems with the brain and spinal cord. Although it has never been proven that folic acid reduces the risk with valproate, a folic acid deficiency is known to cause these same problems in the absence of anticonvulsant drugs.

The same anticonvulsants can also be associated with depletion of vitamin K. Low levels of vitamin K at the time of birth have been associated with an increased risk of bleeding in the baby. Most epilepsy specialists therefore also recommend that women take extra vitamin K, usually ten milligrams per day, during the last month of pregnancy. The newborn baby will often also get an injection of vitamin K shortly after birth.

Breast-feeding

No matter what the drug, experts agree that it is safe for the mother to breast-feed her child if she chooses. Virtually all anticonvulsant drugs will be present in breast milk, but the amount absorbed by the baby will be low and any risk is thought to be much smaller than the benefits of breast-feeding. A possible exception is phenobarbital. Rarely, breast-feeding while the mother is taking this drug can be sedating to the child, sometimes to the point where the baby has difficulty feeding. Mothers taking this drug should simply watch for unusual tiredness in their baby if they choose to breast-feed.

Special Considerations with Small Children

Any parent with epilepsy, no matter how well-controlled, should take extra precautions handling small children, particularly if seizures come with little or no warning. Mostly, this is common sense: any possible risk to the baby should a parent have a seizure needs to be minimized. An example is changing the baby: this should be done on the floor rather than on a table, so that if a seizure occurs, the baby cannot fall from the table. Babies and small children should be bathed on their backs, with only a few inches of water, so that if the child slips into the water, the nose and mouth will remain in the air.

The most important thing to remember about becoming a parent if you have epilepsy is that the vast majority of people with epilepsy have normal, healthy babies. Any pregnancy has risks, but careful planning that takes into account those specific to people with epilepsy will help to further minimize those risks. Above all, people with epilepsy make good parents, and if you want to have children, epilepsy or epilepsy treatment should not prevent you from that goal.

16

Preventing Prejudice: Education Is the Key

Epilepsy has probably been present since early humans roamed the earth; it was mentioned in some of the earliest writings from ancient Egypt. Unfortunately, misunderstanding of this confusing disease has been present just as long. Throughout early history and the Middle Ages, people with epilepsy were generally thought to be possessed by demons—the only explanation people at that time could come up with for their sudden, odd behavior. Physicians only began to widely recognize epilepsy as a disease of the brain in the 1800s, but even then these people were grouped together with the insane. Only during the twentieth century did the medical community recognize epilepsy as a neurologic, rather than a psychiatric, disease. Society lagged behind in this recognition, continuing to view persons with epilepsy as violent or inferior. Until 1965, immigration into the United States was restricted for people with epilepsy. Many states restricted marriage under the mistaken belief that the condition would be inherited, and some even enacted involuntary sterilization laws. The last law forbidding marriage of people with epilepsy was not repealed until 1980, and the last eugenic sterilization law was in effect until 1985.

Although people with epilepsy are now protected by the Americans with Disabilities Act, prejudice and misunderstanding unfortunately still exist. In this chapter we'll explore the major forms of discrimination that people with epilepsy may face, and ways to handle them. Let's begin with an example.

A burly, two-hundred-pound former wrestler who was usually defiant and confident, Nick was in tears in my office. As the droplets rolled down his ruddy cheeks, his mother pushed a newspaper clipping toward me. "They're going to put me away on this one, man," he sobbed. "Ain't nothing I can do about it this time."

Nick was by no means new to trouble with the law, but I instantly understood his frustration and despair this time. I had met him almost ten years before, in the epilepsy clinic after his release from prison. A rough teenager and young adult, he had run with a crowd that turned him to drugs, and he had many, sometimes brutal fights with others in his crowd and with the police. He had blacked out many times, but whether it was from alcohol or from the beating his head took, he couldn't know. Imprisoned for two years for drug possession, he had an opportunity to turn his life around. He stopped using drugs, exercised, and cleared his head. He immediately got a job in construction when he was released, and he stayed away from the old crowd. His mother was beginning to think that her youngest son was now settling down. Then the staring spells started: he seemed "not there" for a minute, for two. Afterward, if anyone asked what had happened, he'd grow angry. "What the f*ck is wrong with you! Nothing happened!" Then, he had a grand mal convulsion, of which he had no memory.

The first time I saw Nick, ten years before, he had been brought (against his will) by his mother. He had no idea why he was there. She thought he had blackouts. "Wouldn't I know if something happened to me?" he demanded. "I don't know why she is making this up!" Even as I spoke with him, though, he paused. His eyes grew glassy, he began to chew. Marjorie nodded.

I held up my watch and pointed at it. "Nick, what is this?" He said nothing. "Say, 'Today is Tuesday.'" He still stared blankly. Then, almost as suddenly, his eyes cleared. "What just happened, Nick?"

"Nothing, man. Why the hell does everybody pick on me?"

"Nick, I only want to . . ." But he was already out, slamming the office door and storming down the hall.

I'd seen people like Nick before. He almost certainly had temporal lobe epilepsy, but he had absolutely no memory of his seizures, such that when he had one, it was as if no time elapsed for him. He also became uncharacteristically agitated, even violent, after seizures. In a few minutes he returned, but he still didn't believe anything had happened.

Over the ensuing years, Nick usually took seizure medication, although he still couldn't believe anything was wrong. A few times he had had seizures in the street. As he walked around, dazed, well-meaning passersby or police sometimes approached. Twice before he had been arrested, either because he was thought to be drunk or because he became agitated when questioned by police. During this time I continued to adjust his medications, and the latest dosage of carbamazepine seemed to be working.

The week before this most recent visit, he'd been driving his van. He hadn't had a seizure that he knew of in almost a year. But, the newspaper clipping said, he had hit a livery cab, then careened into several parked cars before his van finally came to a stop. What had hurt him and his mother the most was the statement, in their hometown paper, that he had been "charged with operating a vehicle under the influence of drugs." His medic alert bracelet had apparently been lying useless in the cab of his car after falling off during the accident and was not seen by the arresting officers, who, no doubt, found him confused, possibly stumbling, and unable to give an account of what had happened because he was in the midst of a seizure.

Law Enforcement

In general, law enforcement officials can be helpful if a person has a seizure in a public place. In most areas, police have basic medical training including aiding people who are having a seizure. They are, however, not doctors. While most police would recognize an obvious grand mal seizure, many would not be able to correctly diagnose the confusion of a complex partial seizure or a postictal state. That's why, as happened to Nick, people with epilepsy sometimes run into trouble.

In people who are well controlled (i.e., seizure-free), it is unlikely that a situation like Nick's would occur. On the other hand, people with frequent seizures are at high risk of having a seizure in a public place, and if they tend to be aggressive afterward, there is more chance of running into trouble. In all cases, the best course is to wear a medic alert bracelet or necklace.

The Workplace

Seizures can be confusing or frightening to coworkers if they are not familiar with them. Like police officers, many people will recognize a grand mal seizure. But if a quiet seizure occurs followed by confused behavior—typically, wandering around, mumbling, or shouting—others could interpret this as willful behavior, drunkenness, or drug use. One patient of mine typically removed his trousers after a seizure—he would say he felt hot—but those unfamiliar with his seizures would of course think this strange behavior.

In dealing with seizures in the workplace, each person must make decisions regarding whether and which coworkers should know about his condition and whether other precautions need to be taken. First, there is the safety of the person should he have a seizure. Second, misunderstandings must be prevented. Finally, the rights of the person with epilepsy must be upheld.

Safety means that the person must not put himself at risk. A person with uncontrolled seizures that impair alertness in any way should not be driving and should not be operating machinery that could place him or others at risk were a seizure to occur. Because of the Americans with Disabilities Act, employers must make reasonable accommodations if these tasks fall into the typical job description. Some people find it helpful if at least a few coworkers (or ideally the supervisor) are aware of their condition. That way, if a seizure occurs, they can help prevent injury, reassure any workers who may not understand the condition, and prevent unnecessary trips to the emergency room. Other people prefer that nobody knows. Typically this is due to embarrassment at their condition, which is unwarranted in most cases. In my experience, people usually find their coworkers supportive, and most of the time (as epilepsy is so common) one or more of them will already have experience through friends or family.

In rare cases, misunderstandings can occur either when the person explains her situation or when a seizure occurs. People unfamiliar with epilepsy may mistakenly believe that she is no longer capable of doing her job, or that her health may be at risk if she continues working. The behaviors associated with seizures can be misinterpreted as laziness, drug use, or insubordination. If a simple explanation does not help, a letter from the treating physician with a brief description of the condition may suffice. Educational materials can also be helpful, such as those provided by the Epilepsy Foundation.

Occasionally, people have lost their job because of epilepsy. This is clearly illegal, and legal counsel should be obtained. The Epilepsy Foundation can also help with this through attorneys on staff who are familiar with this situation. As with any prejudice, education is generally the person's best ally.

Relationships and Marriage

People with epilepsy often have questions about dating and marriage. They may wonder what to tell a boyfriend or a girlfriend, and when. Many of the issues are similar to those of the workplace: in a serious relationship, knowledge of the condition can help to prevent misunderstanding and can also help with maintaining the person's safety. When misunderstanding occurs, education (by the person with epilepsy, the doctor, books, or pamphlets) can be helpful.

I have seen a few cases among specific religious or cultural groups where arranged marriage is the custom and epilepsy is deemed unacceptable. Several of my patients have, for this reason, kept their condition quiet until after the marriage. In general, this is not the best strategy, as the condition virtually always becomes apparent later due to a seizure or to discovery of the medication. Even where epilepsy is unacceptable, however, I have seen education about the condition (particularly the likelihood of normal children) result in acceptance, though this is unfortunately not always the case. Hopefully, with time, this prejudice will become less and less common. In the meantime, people with epilepsy who wish to have an arranged marriage must make individual decisions should problems arise with a diagnosis of epilepsy.

Epilepsy can affect sex drive—almost always to decrease it. This may happen because of certain medications, or independently of them. Medications can also occasionally affect sexual performance. This is an important and frequently overlooked area that should be discussed with your doctor. Medication levels may be too high, or a different medication could be better, or another condition (such as depression or anxiety) may be affecting sexual drive or performance.

I will end with another case, that of the writer Edgar Allan Poe, familiar to all Americans as a brilliant teller of short stories. His chilling tales include "The Pit and the Pendulum" and "The Fall of the House of Usher," stories of madness and terror. Most who know of him immediately think of a life of alcoholism and drug addiction. What most do not realize, however, is that Poe also suffered from epilepsy. In fact, his disease, poorly understood in his day, probably figured in many of his most famous stories where altered consciousness occurred. He is therefore an excellent example of where prejudice and misunderstanding resulted in an erroneous attribution of substance abuse, but also of someone who not only overcame the limitations of his disease but who used insights gleaned from having epilepsy to add to his creative genius.

Edgar Allan Poe suffered from complex partial seizures. In retrospect, it is not difficult to see why these were mistaken for drug addiction. At the time, epilepsy was still a poorly understood disease. The concept of more subtle, "partial" seizures was just beginning to be recognized, so it is not surprising that bizarre behavioral changes would be misconstrued as something else—in this case, abuse of alcohol, opium, and perhaps other drugs. As with many people, his seizures were probably worsened by alcohol, adding to the confusion.

Poe began to suffer from complex partial seizures probably in late adolescence. Perhaps he had the sort of vague, intense emotional experiences (déjà vu) that we now know can be small seizures. Maybe he even had episodes of passing out, dismissed as fainting. He may have had an underlying susceptibility to epilepsy, such that the stress and alcohol consumption of this time in his life caused sufficient irritation of the brain that seizures began. Maybe he had a birth injury—mild trauma with resulting scarring of the brain—that over decades healed but formed connections capable of creating a seizure. Perhaps he was born with a neurotransmitter receptor genetically more efficient than that of most people, making certain brain cells more irritable. Even had he had the advantages of modern testing, the cause of his condition may have remained a mystery (as with many people today).

Medical histories of Poe give brief descriptions of his spells. "When these seizures pass and the person recovers, there may be . . . complete loss of memory. During the attack there is usually loss of self-control and an abnormal ideation." These often occurred after drinking small amounts of alcohol; friends thought that he was merely very susceptible to its effects. But we also know that similar spells occurred in the absence of alcohol. "He looked without seeing, and very frequently became absorbed in visions from which it was very difficult to arouse him." Besides these attacks, Poe was known to have intermittent paranoid and even psychotic behavior. He developed an ongoing paranoia of Longfellow (whom he had at other times described as one of the greatest American poets), accusing him repeatedly of plagiarism. These were probably postictal states, where people remain confused for up to several days after a seizure.

One of Poe's most famous stories is "The Pit and the Pendulum," in which the main character is a man being tortured by the Spanish Inquisition. The beginning, however, offers a description that sounds very much like an epileptic seizure seen through the eyes of a literary genius:

> . . . then, all at once, there came a most deadly nausea over my spirit, and I felt every fiber in my frame thrill as if I had touched the wire of a galvanic battery, while the angel forms became meaningless specters . . . the blackness of darkness supervened; all sensations appeared swallowed up in a mad rushing descent as of the soul into Hades. Then silence, and stillness, and night were the universe.

The narrator then speaks as one who has had such experiences before, implying that this "swoon" was not due entirely to the Inquisition:

He who has [never] swooned, is not he who finds strange palaces and wildly familiar faces in coals that glow; is not he who beholds floating in mid-air the sad perfume of some novel flower; is not he whose brain grows bewildered with the meaning of some musical cadence which has never before arrested his attention.

He seems to indicate some deeper beauty and insight that comes with these events, as have some artists with epilepsy who describe feelings and images from their seizures in painting or writing.

Many people with medical conditions are potentially subject to prejudice, and epilepsy is no exception. The best ally, as in all forms of prejudice, is information and education. If family members, friends, coworkers, and the general public had a better understanding of what epilepsy is—and what it is not—prejudice would not exist. Until then, people with epilepsy, their families, their friends, and their doctors must facilitate that education wherever possible.

17

Finding What Works for You

In any disease it helps to be proactive. With epilepsy you need to work closely with a doctor who understands your particular needs. What are your individual seizures like? How much do they impair your quality of life? How does the medication prescribed make you feel? What are your goals for the next few years, and beyond? These questions are important to you, and only by working together with your physician can you find the answer. Remember that epilepsy is not appendicitis, with one straightforward treatment much the same for everyone. Epilepsy is as individual as the people who suffer from it. If you are not being treated this way, consider seeing another physician or consulting an epilepsy specialist.

In addition to working with your doctor, you need to find what is important to you. (Do you have trouble taking a medicine three times a day? Are you willing to take some risk of a seizure if it means you can try a medicine that is less sedating?) You also need to find what works for you—it may be best for your disease if you always get eight hours of sleep, but it may be important to you to be able sometimes to work long into the night. More than other people, you may feel and function better with a regular exercise program.

You might consider learning stress-reduction techniques, such as yoga, and you should pay attention to your sleep habits so as to maximize your quality of sleep. Again, these factors are very individual but can make the difference between managing your epilepsy and having epilepsy manage you.

Finding the Right Doctor

Epilepsy is a chronic disease, and chances are you will need treatment for at least several years. It is therefore important that your doctor be someone you can work with, not only initially but in whatever situations might arise over the years you are in treatment. In relatively straightforward cases of epilepsy, a primary care physician or general neurologist should be able to care for you. If special situations arise, such as problems with medication, other illnesses, or pregnancy, or certainly if seizures continue, you should consider a consultation with a neurologist who specializes in epilepsy (an epileptologist). He or she can redirect your care, working alongside your general neurologist or primary care physician. In some cases, this may be done in a single visit; in others you may need to see a specialist several times or even regularly (for instance, to guide you through epilepsy surgery). In all cases, however, you should keep in close contact with a local neurologist or your primary case physician as to the course of your treatment.

Working Together with Your Doctor

Doctors in general are well trained in medicine, and particularly in their fields of expertise, but they may sometimes overlook issues that are important to you. If you have concerns about your memory, your mood, family planning, sleep, sex, or any other aspect of your life that may relate to your epilepsy, make sure to bring it up at your visits. Even if the problem is unrelated to epilepsy or your treatment, your doctor may be able to assist you in finding someone who can help.

You also need to take responsibility for your care. If you don't follow your doctor's instructions, it makes it more difficult for your doctor to evaluate the effectiveness of your treatment or to recommend better treatments. If you don't follow instructions for a reason (such as it is too difficult to remember to take medicine several times a day, or taking a full dose makes you sleepy), discuss your concerns with your doctor.

Educate Yourself

It is not possible for your doctor to explain everything you need to know in an office visit, or even several. Even if it were, it is next to impossible for you to remember it all. You have taken a good first step in expanding your knowledge by reading this book. Of course, this is only one outlook on epilepsy, and there are others. Take time to look into important issues. Consult other books if you don't find the answers here, or if you don't think they are right for you. The Internet is becoming an important source of information for all people. You can certainly find a lot of information through the Internet, but do be careful about the source. A chat room, for instance, will offer you a lot of opinions, none of which may be based on fact. People or entities with their own agenda—such as alternative-medicine manufacturers or pharmaceutical companies—may offer information that is in their own best interest rather than yours. Look for reliable, unbiased sources such as universities and national organizations like the Epilepsy Foundation for your information.

Finding Additional Help

Whatever the particular issues, help is available. A good starting point is your local chapter of the Epilepsy Foundation. Physicians who are specialists in epilepsy may be helpful in addressing medical issues, if a primary care physician is not knowledgeable in some aspects of the disease. Many people also find a second or third opinion useful, as no physician knows everything, and there are many different outlooks on particular issues. I encourage this particularly when seizures are very difficult to control. Physicians in psychiatry, obstetrics/gynecology, or other specialties can be helpful as well.

This book is meant to clarify what epilepsy is, but probably more important, what epilepsy is not. In the vast majority of people, epilepsy is not associated with an abnormality in brain structure, and (when not having seizures) the brain typically functions entirely normally. The goal of all treatment should be complete control of seizures with no adverse effects from medication, and this is attainable in the vast majority of cases. Your doctor should agree with this goal and should offer options if it is not met. Even without complete control, however, most people function completely normally the vast majority of the time. With some important limits as we've discussed in previous chapters, people with epilepsy can work, have children, raise a family, and live a full life while they continue to work with their doctor toward full control.

In chapter 16, I mentioned how epilepsy might have influenced the genius of Edgar Allan Poe. History offer many examples of highly talented artists with epilepsy. In some cases, as with Poe and with one of my patients (a painter with visual seizures that influenced his work, mentioned in chapter 2), epilepsy itself may have added to the person's creative abilities. This is also true of Vincent van Gogh, whose wildly impressionistic perceptions of the world were probably influenced by seizures (and, perhaps, by treatments such as digitalis). Fyodor Dostoyevsky incorporated characters with epilepsy into his most famous novels, including *The Idiot* and *The Brothers Karamazov*. Epilepsy, like major depression, perhaps represents not a disease to be feared and shunned, but an extreme of normal. We all become saddened at times, appropriately, and this reaction adds to the breadth and beauty of life. But severe, inappropriate depression can destroy lives and, because of the risk of suicide, is life-threatening and sometimes fatal. Similarly, we all have moments of sudden inspiration, of unexplained thoughts and reactions that can change our lives in positive ways. When these sudden changes become too large or too violent, however, they are called seizures. Rather than adding inspiration, they impair our lives and can destroy them. Keeping this perspective on epilepsy will help to alleviate the fear and prejudice that exist in the world and allow more people with epilepsy to lead the full, productive lives of the people described in this book.

In confronting any disease or indeed any adversity, planning, perspective, and goals are critical. With these in mind, all persons can live well with epilepsy.

Appendix: Sources of Further Information

General Information

Epilepsy Foundation: www.epilepsyfoundation.org
4351 Garden City Drive
Landover, MD 20785-7223
(800) 332-1000

The best resource for general, patient-oriented information and services related to all aspects of epilepsy. The Web site also provides links to regional epilepsy foundations.

Epilepsy Foundation Local Chapters

Up to date addresses can also be obtained from the national office, above.

Alabama

Epilepsy Foundation of North and Central Alabama
701 Thirty-seventh St. South, Suite 8
Birmingham, AL 35222-3222
(205) 324-4222
(800) 950-6662

Epilepsy Foundation of South Alabama
951 Government St., Suite 201
Mobile, AL 36604-2425
(251) 432-0970
(800) 626-1582

Arizona

Epilepsy Foundation of Arizona
PO Box 25084
Phoenix, AZ 85002-5084
(602) 406-3581
(888) 768-2690

California

Epilepsy Foundation of Los Angeles, Orange, San Bernardino & Ventura Counties
5777 West Century Blvd., Suite 820
Los Angeles, CA 90045-5664
(310) 670-2870
(800) 564-0445

Epilepsy Foundation of Northern California
1624 Franklin St., Suite 900
Oakland, CA 94612-2824
(510) 893-6272
(800) 632-3532

Epilepsy Foundation of San Diego County
2055 El Cajon Blvd.
San Diego, CA 92104-1091
(619) 296-0161

Colorado

Epilepsy Foundation of Colorado, Inc.
234 Columbine St., Suite 333
Denver, CO 80206-4711
(303) 377-9774
(888) 378-9779

Connecticut

Epilepsy Foundation of Connecticut, Inc.
386 Main St.
Middletown, CT 06457-3360
(860) 346-1924
(800) 899-3745

Delaware

Epilepsy Foundation of Delaware
Tower Office Park
Newport, DE 19804-3167
(302) 999-9313
(800) 422-3653

District of Columbia

Epilepsy Foundation of the Chesapeake Region
Hampton Plaza
Towson, MD 21286-3012
(410) 828-7700
(800) 492-2523

Florida

The Epilepsy Foundation of Northeast Florida, Inc.
5209 San Jose Blvd., Suite 101
Jacksonville, FL 32207-7663
(904) 731-3752
(888) 897-8579

Epilepsy Foundation of South Florida, Inc.
7300 North Kendall Dr., Suite 700
Miami, FL 33156-7840
(305) 670-4949

Georgia

Epilepsy Foundation of Georgia
100 Edgewood Ave. NE, Suite 806
Atlanta, GA 30303-3067
(404) 527-7155
(800) 527-7105

Hawaii

Epilepsy Foundation of Hawaii, Inc.
245 North Kukui St., Suite 207
Honolulu, HI 96817-3921
(808) 528-3058

Idaho

Epilepsy Foundation of Idaho
310 West Idaho
Boise, ID 83702-6039
(208) 344-4340
(800) 237-6676

Illinois

Epilepsy Foundation of Greater Chicago
20 East Jackson Blvd., Suite 1300
Chicago, IL 60604-2248
(312) 939-8622
(800) 273-6027

Epilepsy Foundation of North/Central Illinois
321 West State Street, Suite 208
Rockford, IL 61101-1119
(815) 964-2689
(800) 221-2689

Epilepsy Foundation of Southwestern Illinois
1931 West Main St.
Belleville, IL 62226-7479
(618) 236-2181

Indiana

Epilepsy Foundation Kentuckiana
501 East Broadway, Suite 110
Louisville, KY 40202-1797
(502) 584-8817
(866) 275-1078

Kansas

Epilepsy Foundation of Kansas & Western Missouri
6550 Troost Ave., Suite B
Kansas City, MO 64131-1266
(816) 444-2800
(800) 972-5163

Kentucky

Epilepsy Foundation Kentuckiana
501 East Broadway, Suite 110
Louisville, KY 40202-1797
(502) 584-8817
(866) 275-1078

Louisiana

Epilepsy Foundation S.E. Louisiana
3701 Canal Street, Suite H
New Orleans, LA 70119-6101
(504) 486-6326
(800) 960-0587

Maryland

Epilepsy Foundation of the Chesapeake Region
Hampton Plaza
Towson, MD 21286-3012
(410) 828-7700
(800) 492-2523

Massachusetts

Epilepsy Foundation of Massachusetts & Rhode Island
540 Gallivan Blvd., 2nd Floor
Boston, MA 02124-5401
(617) 506-6041
(888) 576-9996

Michigan

Epilepsy Foundation of Michigan
20300 Civic Center Dr.
Southfield, MI 48076-4105
(248) 351-7979
(800) 377-6226

Minnesota

Epilepsy Foundation of Minnesota
2356 University Ave. West, Suite 405
Saint Paul, MN 55114-1850
(651) 646-8675
(800) 779-0777

Mississippi

Epilepsy Foundation of Mississippi
PO Box 16232
Jackson, MS 39236-6232
(601) 362-2761
(800) 898-0291

Missouri

Epilepsy Foundation of the St. Louis Region
7100 Oakland Ave.
Saint Louis, MO 63117-1813
(314) 645-6969
(800) 264-6970

New Jersey

Epilepsy Foundation of New Jersey
429 River View Plaza
Trenton, NJ 08611-3420
(609) 392-4900
(800) 336-5843

New York

Epilepsy Foundation of Long Island, Inc.
506 Stewart Ave.
Garden City, NY 11530-4706
(516) 739-7733
(888) 672-7154

Epilepsy Foundation of New York City
305 Seventh Ave., 12th Floor
New York, NY 10001-6008
(212) 633-2930

Epilepsy Foundation of Northeastern NY
Three Washington Square
Albany, NY 12205-5523
(518) 456-7501
(800) 894-3223

Epilepsy Foundation of Rochester-Syracuse-Binghamton
1650 South Ave., Suite 300
Rochester, NY 14620-3901
(716) 442-4430
(800) 724-7930

Epilepsy Foundation of Southern New York
One Blue Hill Plaza, Box 1745
Pearl River, NY 10965-3104
(845) 627-0627
(800) 640-0371

North Carolina

Epilepsy Foundation of North Carolina, Inc.
3001 Spring Forest Rd.
Raleigh, NC 27616-2817
(919) 876-7788
(800) 451-0694

Ohio

Epilepsy Foundation of Central Ohio
454 East Main St., Suite 250
Columbus, OH 43215-5354
(614) 228-4401
(800) 878-3226

Epilepsy Foundation of Greater Cincinnati, Inc.
895 Central Ave., Suite 550
Cincinnati, OH 45202-5757
(513) 721-2905

Epilepsy Foundation of Northeast Ohio
2800 Euclid Ave., Suite 450
Cleveland, OH 44115-2418
(216) 579-1330

Epilepsy Foundation of Northwest Ohio
1251 South Reynolds Rd., PMB 323
Toledo, OH 43615-6938
(248) 351-7979
(800) 377-6226

Epilepsy Foundation of Western Ohio
7523 Brandt Park
Huber Heights, OH 45424-2337
(937) 233-2500
(800) 360-3296

Oregon

Epilepsy Foundation of Oregon
619 SW Eleventh Ave., Suite 225
Portland, OR 97205-2646
(503) 228-7651
(888) 828-7651

Pennsylvania

Epilepsy Foundation of Eastern Pennsylvania
919 Walnut St., Suite 700
Philadelphia, PA 19107-5237
(215) 629-5003
(800) 887-7165

Epilepsy Foundation of Western & Central Pennsylvania
1323 Forbes Ave., Suite 102
Pittsburgh, PA 15219-4725
(412) 261-5880
(800) 361-5885

Puerto Rico

Sociedad Puertorriqueña de Epilepsia
Hospital Ruiz Soler
Bayamón, PR 00959
(787) 782-6200
(787) 782-6262

Rhode Island

Epilepsy Foundation of Massachusetts & Rhode Island
540 Gallivan Blvd., 2nd Floor
Boston, MA 02124-5401
(617) 506-6041
(888) 576-9996

South Carolina

Epilepsy Foundation of South Carolina
652 Bush River Rd., Suite 211
Columbia, SC 29210-7537
(803) 798-8502

Tennessee

Epilepsy Foundation of East Tennessee
PO Box 3156
Knoxville, TN 37927-3156
(865) 522-4991
(800) 522-4991

Epilepsy Foundation of Middle Tennessee
2002 Richard Jones Rd., Suite C202
Nashville, TN 37215-2809
(615) 269-7091
(800) 244-0768

Epilepsy Foundation of Southeast Tennessee
744 Mccallie Ave., Suite 517
Chattanooga, TN 37403-2520
(423) 756-1771

Epilepsy Foundation of West Tennessee
P.O. Box 382610
Germantown, TN 38183-2610
(901) 854-0114

Texas

Epilepsy Foundation of Central & South Texas
10615 Perrin Beitel Rd., Suite 602
San Antonio, TX 78217-3142
(210) 653-5353
(888) 606-5353

Epilepsy Foundation of Greater North Texas
2906 Swiss Ave.
Dallas, TX 75204-5929
(214) 823-8809

Epilepsy Foundation of Southeast Texas
2650 Fountain View Dr., Suite 316
Houston, TX 77057-7619
(713) 789-6295
(888) 548-9716

Vermont

Epilepsy Foundation of Vermont
PO Box 6292
Rutland, VT 05702-6292
(802) 775-1686

Virginia

Epilepsy Foundation of the Chesapeake Region
Hampton Plaza
Towson, MD 21286-3012
(410) 828-7700
(800) 492-2523

Epilepsy Foundation of Virginia
UVA Health Services Center
Charlottesville, VA 22908-0001
(434) 924-8669

Washington

Epilepsy Foundation Washington
3800 Aurora Ave. North
Seattle, WA 98103-8798
(206) 547-4551
(800) 752-3509

West Virginia

Epilepsy Foundation of West Virginia
238 Fourth Ave., Suite 106
South Charleston, WV 25303-1539
(304) 746-9570
(866) 746-9570

Wisconsin

Epilepsy Foundation of Central & Northeast Wisconsin
1004 First St., Suite 5
Stevens Point, WI 54481-2627
(715) 341-5811
(800) 924-9932

Epilepsy Foundation South Central Wisconsin
1302 Mendota St.
Madison, WI 53714-1024
(608) 442-5555
(800) 657-4929

Epilepsy Foundation of Southeast Wisconsin
735 North Water St., Suite 701
Milwaukee, WI 53202-4104
(414) 271-0110

Epilepsy Foundation of Southern Wisconsin, Inc.
205 North Main St., Suite 106
Janesville, WI 53545-3062
(608) 755-1821
(800) 693-2287

Epilepsy Foundation of Western Wisconsin
1812 Brackett Ave., Suite 5
Eau Claire, WI 54701-4677
(715) 834-4455
(800) 924-2105

There are no local chapters for the following states:

Alaska	New Hampshire
Arkansas	New Mexico
Iowa	North Dakota
Maine	South Dakota
Montana	Utah
Nebraska	Wyoming
Nevada	

Other National Organizations

National Institute of Neurological Diseases and Stroke:
www.ninds.nih.gov
PO Box 5801
Bethesda, MD 20824
(800) 352-9424

This site provides general information about epilepsy as well as specific descriptions of ongoing research at the National Institutes of Health.

American Epilepsy Society: www.aesnet.org
324 North Main St.
West Hartford, CT 06117
(860) 586-7505

This organization is mainly for health care professionals; however, it offers abundant information about epilepsy that is useful for patients as well. In particular, this Web site has information about patient-assistance programs for lower-income patients who cannot afford their prescription drugs.

American Academy of Neurology: www.aan.com
1080 Montreal Ave.
Saint Paul, MN 55116
(651) 695-1940

Child Neurology Society: www.childneurologysociety.org
1000 West County Rd. East, Suite 290
Saint Paul, MN 55126
(651) 486-6688

Society for Neuroscience: www.sfn.org
11 Dupont Circle
Washington, DC 20036
(202) 462-6688

This organization is mainly for scientists interested in all aspects of the brain and brain disease.

Citizens United for Research in Epilepsy (CURE)
8110 Woodside Lane
Burr Ridge, IL 60525
(630) 734-9957

This organization works toward encouraging research in pediatric epilepsy, mainly through private funding.

Epilepsy Institute
257 Park Ave. South
New York, NY 10010
website@epilepsyinstitute.org
www.epilepsyinstitute.org
Tel: 212-677-8550
Fax: 212-677-5825

Pregnancy Registry
1-888-233-2334

The goal of this organization is to learn more about the effects of epilepsy drugs on children exposed to them. Pregnant women who call this number will be asked some questions regarding their health, and after the child is born, they will be contacted again. This is the best way we now have to sort out the risks of epilepsy drugs in pregnancy, particularly the risks of the newer drugs.

Patient Prescription-Assistance Plans

www.aesnet.org—Patient drug-assistance plans are listed on this Web site of the American Epilepsy Society.

www.needymeds.com—This site provides information on patient-assistance programs from all pharmaceutical companies. Some companies have applications that can be downloaded on your computer.

www.together-rx.com—This site offers a discount card for Medicare recipients and is provided by seven different companies. The card can be used to receive discounts on over 150 different medications. Limited income is a requirement.

Safety of Herbal Products

National Center for Complementary and Alternative Medicine:
http://altmed.od.nih.gov

Office of Dietary Supplements:
http://dietary-supplements.info.nih.gov

National Institute of Medical Herbalists
56 Longbrook St.
Exeter, United Kingdom EX4 6AH
nimh@ukexeter.freeserve.co.uk
ww.nimh.org.uk

Register of Chinese Herbal Medicine: www.rchm.co.uk

Index

Page numbers in *italics* refer to figures and tables.

insulinomas, 79
intelligence, 31, 41
interictal images, 145
Internet, 223
internists, 80, 81
intracarotid amobarbital test, *see* Wada test
intracranial recording, 90–91, 142
ion channels, 39
IUDs, 202

Jackson, Hughlings, 17
Jacksonian march, 17, 19
jerking, 16, 17, 18–19, 39
JME, *see* juvenile myoclonic epilepsy
Joel, Billy, 13, 24, 155, 156
jogging, 151
Joyner, Al, 190, 191
Joyner, Florence Griffith "Flo Jo," 189–91
juvenile myoclonic epilepsy (JME), 29–30, 31, 39, 150, 152–53, 206

kava, 164
Keppra, *see* levetiracetam
ketogenic diet, 160–61
ketones, 160, 161
kidneys, 61, 209
kidney stones, 99, 116, 118, 160
Klonopin (clonazepam), 153, 172

lamotrigine (Lamictal), 29, 30, 47, 56, 61–62, 63, 97, 98, 105, 112, 115–16, 117, 121, *124*, 172, 195, 196, 201–2, *203*

language, 99, 130–31, 135, 144
and the brain, 11–12, 18, 85, 88, 89, 91–92, 133–35
seizures and, 35–36, 37–38
in temporal lobe seizures, 22–23, 139, 142
laser surgery, 178
law enforcement personnel, 192, 215
lead, 80
learning, 85, 98, 103
Lennox-Gastaut syndrome, 30–31
levetiracetam (Keppra), 62, 97–98, 99, 117, *125,* 153, 171, 195, *203*
Lexapro (escitalopram), 194
libido, 166
lips, biting of, 181
liver, 61, 105, 164, 165, 209
liver disease, 99, 108, 113, 167
liver enzymes, 62
liver tests, 108
localized resection, 132, 133, 144, 145
local seizures, *55*
Longfellow, Henry Wadsworth, 219
lorazepam, 153
Lou Gehrig's disease, 128
lumbar puncture, 79
lungs, 189
Lyrica (pregabalin), 118–19

magnesium, 79
magnetic resonance imaging, *see* MRI
magnetic stimulation, 177
major motor seizures, 18–20, *33*